AF765622

AN ILLUSTRATED KEY TO

Freshwater and Soil Amoebae

with notes on Cultivation and Ecology

by

F. C. PAGE

Culture Centre of Algae and Protozoa, Cambridge
Institute of Terrestrial Ecology

FRESHWATER BIOLOGICAL ASSOCIATION
SCIENTIFIC PUBLICATION No. 34

DURHAM UNIVERSITY
SCIENCE
27 NOV 1976
SECTION
LIBRARY

PREFACE

Amoeba is the animal with which many of us were first introduced to biology. It is therefore surprising that the taxonomy and general biology of amoebae are poorly known to all but a few experts. This key is an attempt to help put this right. It is also a departure for the series because it is illustrated largely by photographs; as explained by Dr Page the nature of amoebae really demands this. In printing the photographs we have attempted both to show adequate detail and yet retain a modest price for the key. Inevitably, however, it costs rather more than other recent numbers in the series.

The Association has been fortunate to have the cooperation of Dr Page and his Institute in the preparation of this publication and hopes that it will stimulate more biologists to initiate studies on amoebae in the natural environment.

THE FERRY HOUSE
August 1976

E. D. LE CREN
Director

593.12
PAG

SBN 900386 26 6

ISSN 0367-1887

CONTENTS

INTRODUCTION

Of all the amoeboid cells and organisms among protistans, this key covers one rather restricted group, which is nevertheless the group ordinarily intended by the term 'amoeba.' These are the protozoa classified in the subclass Gymnamoebia (see classification on pages 145–147), a group approximately equivalent to the order Amoebida (Amoebina of the older literature) in the former, broader sense (Honigberg *et al.*, 1964). The relevant classes, subclasses, orders, and sub-orders are defined in the formal classification on pages 145–147. Since this is intended to be a practical manual, the reader is referred to Page (1976) for the theoretical background of the classification. It may be stated here that even in its more restricted sense the term Amoebida still includes most organisms comprehended in the older and broader definition.

Members of the subclass Gymnamoebia carry on both locomotion and feeding by protoplasmic streaming, with or without discrete pseudopodia. The usual statement that they move by pseudopodia applies only if one considers the advancing tip a pseudopodium, whether or not it projects distinctly from the main mass. The order Amoebida in its former sense (equivalent to the present subclass Gymnamoebia) has usually been described as comprising *lobose* amoebae, that is, with blunt rather than finely pointed pseudopodia, but many organisms with fine sub-pseudopodial projections differ in no other discernible way from other members of the subclass. Such sub-pseudopodia are discussed in detail later. Axopodia or reticulopodia do exclude an organism from the Gymnamoebia, and many with filopodia do not belong here for one reason or another and are excluded by the definitions of the relevant taxa.

Amoebae in this sense are heterotrophic and not members of any group (lower than phylum) containing autotrophs. Most lack any flagellate stage, though a few have a more or less transient flagellate phase. Most are uninucleate; a few are binucleate; the few multinucleate species appear to reach that condition by fission into multinucleate daughter cells or nuclear division within originally uninucleate cells, but not by fusion. Gymnamoebia do not produce spores or fruiting bodies, though many produce resting cysts. In many parasitic and at least two free-living species, nuclear division occurs within the cyst. In one free-living species, there is almost certainly a sexual process in the cyst.

Finally, the Gymnamoebia are naked amoebae, that is, lacking a test (shell). This distinction is somewhat blurred not only by the Cochliopodiidae, whose flexible coating makes them disputed territory between Gymnamoebia and Testacialobosia (lobose amoebae with tests) but also by the finding of scales, indistinguishable with the light microscope, covering the entire surface of some unquestioned Gymnamoebia, namely certain species of *Mayorella* and *Paramoeba*.

Other organisms found in freshwater and soil habitats but not included in this key are cellular and plasmodial slime moulds (see the excellent taxonomic summaries of Olive (1975) and Raper (1973)) and members of that rather hotch-potch group the proteomyxans. A brief section on the Leptomyxida is included, separate from the key to the Gymnamoebia, since the general investigator may be puzzled by these organisms and unable to find a hint of their identity elsewhere.

No organisms that are ordinarily parasitic are included, e.g. *Hydramoeba* and *Entamoeba* (except *E. moshkovskii*), but the sections on *Naegleria* and *Acanthamoeba*, ordinarily free-living but sometimes invading man and other animals, contain information meant to be helpful to workers interested in those genera primarily as parasites and pathogens. Nor are genera with species only in salt water included, though two of those genera, *Paramoeba* and *Flabellula*, give their names to families included in the key.

One of the principal aims has been to sort out well-based genera and clearly described species from the mass of names and vague descriptions in the literature. Species are included if the original or subsequent published descriptions seem clear, or if new observations make it possible to sharpen the diagnosis. The purpose has been to make the key as complete and usable as possible, not to produce a catalogue of names. For some genera, there are many more named species than are included. It has not been possible even to include all species named in recent years, because some descriptions seemed inadequate. The non-specialist should be warned that there is considerable disagreement on genera among students of the Gymnamoebia. He should also expect to meet undescribed species, even in common genera such as *Thecamoeba* and *Vannella*. Nomenclatural innovations on the specific level (new species, changes in names) have been held to a minimum. Apart from transfers between genera, the principal changes in names have been corrections of adjectival endings; few if any authors other than Lepşi seem to have noticed that *Chaos* is a neuter noun.

Amoebae are much more abundant and widely distributed than is generally appreciated. It is true that one may examine many drops from a collected water sample without seeing (or at least noticing) an amoeba, but in fifteen years I have never inoculated fresh or salt water or soil on to agar without having amoebae grow out. Too little is known about their geographical and ecological range. Some species have been found in many parts of the globe. Although certain genera have often been called 'soil amoebae', this designation means only that they occur in terrestrial as well as aquatic habitats, while some other groups have never been found in the soil, narrowly defined. It should be noted that among those amoebae found in soil and leaf litter are well-studied species for which no cyst is known, as well as cyst-forming species. The relationship between freshwater and marine amoebae has not yet been completely worked out. Although one species has been reported from both salt and fresh water on the basis of morphological similarity, present evidence suggests that most species occur in only one or the other of those environments. Some genera occur in both fresh and salt water, while others appear exclusively marine.

A consideration of habitats would not be complete without mentioning that some 'normally' free-living species have now been found to be facultative invaders of the bodies of multicellular animals and, in one case, plants. The species principally involved in endozoic situations, so far as we now know, belong to the two genera previously mentioned, *Naegleria* and *Acanthamoeba*, though it is quite likely that others may be found to be at least harmless visitors. The term *amphizoic* has been proposed as a general designation for protozoa which may be either free-living or endozoic, whichever environment is the 'normal' one.

There is no geographic limitation on the species included in this key, just as there appears to be no narrow geographic limit to the distribution of many species. The notes on places from which certain species have been reported are not meant to suggest that they do not occur elsewhere. The disproportionate number of times that an American habitat is mentioned does not indicate an unusually large number of species in that continent.

METHODS

COLLECTION AND EXAMINATION OF FIELD MATERIAL

Unless one intends to disregard all cyst-forming amoebae which may grow out, it is necessary to collect with sterile apparatus. Ordinary jam jars are quite satisfactory for much collecting if one takes care not to contaminate with one's fingers the material being collected. To avoid stirring up the bottom of bodies of water or to reach less accessible spots, a simple collecting device may be made: a 150 ml Erlenmeyer flask with a bung, through which are inserted one bent glass tube (*c.* 60°) for connecting to a longer straight tube, and another bent glass tube (*c.* 120°) for aspirating material into the flask by mouth. A short rubber or flexible plastic tube connects a long straight tube with one bent tube leading into the flask. At the point where this connection is made the apparatus can be taken apart for autoclaving, with all open ends closed by means of aluminium foil; the two parts are quickly assembled at the collection site. The long straight tube is dipped to the spot from which material is desired. The collected material may be poured quickly into a sterile jar for carrying.

Superficial soil and leaf litter samples can also be collected with small jars; if these are autoclaved, a covering of aluminium foil may first be applied to the upper part of the jar to ensure sterility of the edge used for scraping the ground. If any sort of coring device is used, attention must again be given to sterility.

For quantitative collecting methods, the reader is referred to works on hydrobiology (e.g., Rodina, 1972) and soil biology (e.g., Heal, 1971; Parkinson *et al.*, 1971; Singh, 1955, 1975).

One should search the collected material for amoebae as soon as possible, since some will not grow out by the methods given, but most amoebae present will be missed unless the field material is cultured out. After inoculating a little into culture vessels (see following sections), one can examine the rest without further concern about contamination. Let the collected material settle, then put a drop taken from the bottom on to each of two or three slides and add to each drop a coverslip with a narrow streak of petroleum jelly along two parallel edges of the downward side. It may be helpful also to prepare hanging drops, as described on p. 22, though these are not always as necessary at this stage as when examining bacteria-rich cultures.

Although details on observation are given in the later section on that topic, suggestions especially applicable to observations on flat-side preparations will be found useful at this stage. The petroleum jelly will prevent the coverslip from crushing any larger amoebae, but care must be taken not to let the preparation dry out through the two sides left open for entry of oxygen. The intensity of the illumination should be reduced to a fairly low level, and a search for amoebae should be conducted with a magnification of approximately ×100. Give a few minutes for the amoebae to extend into the normal locomotive form. Some will retain the floating form for hours, and others will round up and never begin locomotion under these conditions. These problems are discussed in the later section on observation. Nevertheless, it will be possible to observe locomotive forms of some species in the relatively bacteria-poor preparations from field material.

The main purpose of such examinations is to record the species present. However, if large amoebae (species of *Amoeba*, *Chaos*, *Pelomyxa*, even the larger *Mayorella* and *Thecamoeba*) are found on these slides, the lack of aseptic precautions should not prevent attempts to culture them, since the possibility of contamination is nil, at least with the first three genera mentioned. The slide and coverslip may be separated carefully and placed into one of the liquid media recommended in the section on cultivation. Unfortunately, success in starting a culture of a large amoeba from a single individual on a slide is rather rare.

PRINCIPLES OF CULTIVATION

The purposes of cultivation are enrichment and isolation. By enrichment, using culture conditions favouring those organisms that one is seeking, one detects species that would be missed by direct examination of the collected material. By isolation one obtains a culture of organisms of a single species, that is, a population. This definition must be modified in the case of amoebae and most other protozoa to recognise that organisms of other species are usually present as food, but no organisms of a similar species, likely to be confused with the object of study, are present. Usually one ensures that all the organisms thought to belong to a single species actually do, by isolation of clones, a procedure described later.

For organisms to survive and multiply they must eat. Fortunately, many amoebae do very well on the bacteria accompanying them from nature, and upon subculturing on to a weakly nutrient medium these food organisms will also multiply. One can also add bacteria of an identified strain, such as *Klebsiella pneumoniae* or *Escherichia coli*.

Some amoebae seem unable to live on bacteria alone. As might be expected, these are medium-sized and larger amoebae. Some species of *Thecamoeba* ingest and multiply better on small amoebae and bacteria than on bacteria alone. A small *Vannella* has been fed to *T. sphaeronucleolus*; *Naegleria gruberi* to *T. striata*; and a small *Vexillifera* to *T. quadrilineata*. At least some species of *Mayorella* seem to require smaller protozoa or algae. In our cultures *M. riparia* feeds on the flagellate *Rhynchomonas nasuta*, and *M. penardi* on the amoeba *Vexillifera* sp. The colourless flagellate *Chilomonas paramecium* has long been used as food for *Amoeba proteus* and also serves for *Thecamoeba proteoides*. The ciliates *Tetrahymena*, *Colpidium*, and *Paramecium* are standard laboratory diet for *A. proteus* and *Chaos carolinense*. Filamentous blue-green algae have been used as food for one species of *Thecamoeba*, and green flagellates and diatoms for other amoebae. Although a long-term culture method for *Pelomyxa palustris* is still lacking, these large amoebae do ingest blue-green filaments.

The foregoing should not be taken as suggesting that one cannot hope to culture amoebae without first obtaining a wide supply of possible food organisms. In most cases, some of the food organisms will be collected with the predator. Obviously, if an amoeba grows out in mixed cultures made from collected material, one or more food species are present. One then needs only to provide for further growth of the prey organism, and most of the culture media suggested in later sections do support the food organisms. It is advisable to have on hand cultures of the food bacteria there mentioned if one wishes to cultivate amoebae on non-nutrient agar. These food bacteria can be grown on nutrient agar slopes. If one wishes to grow *Amoeba proteus* or *Chaos carolinense* (whether one is fortunate enough to isolate these or has to acquire them from an existing strain), it may be necessary to get cultures of one or two food organisms.

Where there is a choice between growing amoebae on bacteria or on algae, consideration should be given to the facts that cultures feeding on algal cells are freer of particulate matter and probably have a lower level of toxic products, but cultures feeding on bacteria generally become denser.

This is not the place to go into axenic cultivation, i.e., cultivation in the complete absence of any organisms but those of the one desired species. Relatively few amoebae have been brought into axenic culture, though the number undoubtedly will increase.

CULTURE MEDIA

There is no single suitable medium for all amoebae that one might encounter in a survey of a given environment. A medium must be suitable for both the amoebae being cultivated and the prey organisms, and it must result in sufficient but not excessive growth of the prey, unless the latter is being added to a non-nutrient medium. The purpose of this section is to give the compositions of some useful media and to indicate those appropriate for given organisms. These media are intended for long-term maintenance as well as for the initial cultivation from the wild.

Whether the final medium is liquid or solid, it has a liquid base. Even tap water may succeed if its use is unavoidable, but it can lead to problems. In some cases relatively unpolluted natural waters may give better results than the usual laboratory solutions. However, natural waters must be filtered immediately after collection, and autoclaving of either tap water or natural waters may result in formation of precipitates.

Saline solutions

Of the large number of very weak saline solutions in use, three are of wide application.

Modified Neff's amoeba saline (*AS*) (Page, 1967): Make up a separate stock solution of each of the following components by dissolving the amount given in 100 ml of glass-distilled water. Prepare the final dilution by adding 10 ml of each stock solution to enough glass-distilled water to make 1 litre.

NaCl	1·20 g
$MgSO_4.7H_2O$	0·04 g
$CaCl_2.2H_2O$	0·04 g
Na_2HPO_4	1·42 g
KH_2PO_4	1·36 g

Prescott's and James's solution (*PJ*) (Prescott & James, 1955): Make up the three stock solutions, each with 100 ml of glass-distilled water. The final dilution consists of 1 ml of each stock solution and 997 ml of glass-distilled water.

Stock solution A

$CaCl_2.2H_2O$	0·433 g
KCl	0·162 g

Stock solution B

K_2HPO_4	0·512 g

Stock solution C

$MgSO_4.7H_2O$	0·280 g

Chapman-Andresen's modified Pringsheim's solution (Chapman-Andresen, 1962): Make up a separate stock solution of each of the following components in 100 ml of glass-distilled water. The final dilution is made by adding 1 ml of each stock solution to 995 ml of glass-distilled water.

$Ca(NO_3)_2.4H_2O$	20·0 g
$MgSO_4.7H_2O$	2·0 g
$Na_2HPO_4.2H_2O$	2·0 g
KCl	2·6 g
$FeSO_4.7H_2O$	0·2 g

LIQUID MEDIA

Most of these are infusions or soil extracts. All except HSM should be used in small covered petri or evaporating dishes, 5–10 cm in diameter, with the fluid 7–10 mm deep.

Grass-seed infusion (*GS*): Boil a few seeds, preferably untreated with fungicides or pesticides, in water or a saline solution such as PJ for 5 minutes. A good proportion is 2 g grass seed to 1 litre fluid. However, no more than 6–12 grass seeds should be left in a dish, depending on size of the dish. If there is time, leave standing lightly covered for a day or two before inoculating.

Cerophyl-Prescott infusion (*CP*): Boil 1 g of the cereal derivative Cerophyl (Cerophyl Laboratories Inc., Kansas City, Missouri, U.S.A.) in 1 litre of PJ solution for 5 minutes, filter out the particles, and restore the volume with distilled water.

Rice-grain infusion: Pour PJ fluid or Chapman-Andresen's modified Pringsheim's solution to a depth of 1 cm or slightly more in a 9- or 10-cm diameter evaporating dish, and add three or four *unboiled* rice grains. Let stand for a day or two before inoculating.

A *soil-extract medium* useful for amoebae feeding on diatoms and other algae is one originated by Professor E. G. Pringsheim and known in our laboratory as E+S. Mix equal quantities of up to three soils (preferably good calcareous garden loam). Into a beaker put enough soil mixture and tap water so that the supernatant water occupies approximately three-fifths of the combined depth. Cover, and steam for one hour on each of two successive days. Decant or filter the liquid, which is the soil extract here designated as E. E+S liquid consists of:

E	10 ml
K_2HPO_4, 0·1% w/v	2 ml
$MgSO_4.7H_2O$, 0·1% w/v	2 ml
KNO_3, 1·0% w/v	2 ml
Glass-distilled water	84 ml

The *HSM medium* devised by W. R. Jones (1946) and used by R. A. Neal (1953) for the cultivation of *Entamoeba moshkovskii* in tubes contains 0·5 ml sterile horse serum, 1 ml 1% Marmite solution, 8·5 ml of a buffered saline solution, and 30 mg rice starch. Jones gave the composition of the buffer solution (pH 7·2) as

Na_2HPO_4 (9·476 g/l)	375 ml
KH_2PO_4 (9·078 g/l)	125 ml
NaCl (0·9% solution)	2250 ml

Agar media

Agar media have been in use for cultivation of amoebae for at least 80 years, but Singh (1945) pioneered the wide use of non-nutrient agar (avoiding excessive bacterial populations) in recent years. The agar media given here are made with non-nutrient agar, such as Difco Bacto-Agar or Oxoid Agar No. 4.

Grass-seed agar: To 1 litre hot GS infusion made as described above (preferably with AS or PJ liquid) add 15 g non-nutrient agar and dissolve. The autoclaved medium is dispensed into sterile petri dishes to a depth of 5 mm or less, swirling the flask to get *a few* seeds into each dish. Discard the excess seeds.

Cerophyl-Prescott agar: Made with 1 litre of the CP infusion described above and 15 g of non-nutrient agar.

Non-nutrient agar (NN): Made with 1 litre of amoeba saline (AS) and 15 g of non-nutrient agar. Must be streaked with bacteria (*Klebsiella pneumoniae* or *Escherichia coli*) before inoculation with amoebae.

Stoianovitch's malt extract–yeast extract agar (MYAS): Made with 0·1 g malt extract, 0·1 g yeast extract, 15 g non-nutrient agar, and 1 litre AS.

Napolitano's agar: 0·2 g beef extract, 0·02 g yeast extract, 6 g glucose, 5 g sodium chloride, 10 g non-nutrient agar, and glass-distilled water to make 1 litre.

All agar media should be sterilized and dispensed into sterile petri dishes. Amoebae are transferred from one agar surface to another by making a cell suspension in the appropriate fluid (e.g., AS or PJ), rinsing the culture surface with a sterile Pasteur pipette. I have found it convenient to sterilize and dispense the transfer fluid from test tubes containing about 3 ml of fluid and stoppered with cotton wool.

Other media, liquid and agar, are in use. For example, Bark (1973) found a medium made from soil extract, Complan and dried hard-boiled egg yolk most suitable for some *Cochliopodium*. The reader is encouraged to devise his own variations.

Selection of media

It should be emphasized that, unfortunately, no single medium is adequate for all amoebae or for all stages of investigation. Agar media are excellent for many medium-sized and most small amoebae, and they have the additional advantage that when inoculated with mixed field material they help to eliminate many of the accompanying organisms which need more fluid than the thin film on the agar surface. Unless otherwise indicated (as in the next section) recommendation of an agar medium implies that no visible liquid overlay is left on the surface. Free liquid on a bacterized agar surface actually inhibits growth of many amoebae, strange though that may seem for creatures of aquatic origin. The reason is probably the accumulation of bacterial wastes and reduction of oxygen concentration. On the other hand, the largest amoebae will usually not survive and multiply on such an agar surface and require more fluid.

If one is trying to grow out all the amoebae present in a given habitat, one will need a range of media to start with. This problem is considered in the next section. At this point we are concerned with the preferred medium for each group of amoebae once they are in culture.

Amoeba proteus: Rice-grain infusion with PJ, inoculated with *Chilomonas paramecium*. The flagellate itself thrives in this medium.

Chaos carolinense: Since this organism is rare, few investigators will face the problem of growing it out from field material, but just because of its rarity, this and other species of *Chaos* are of special interest, and an attempt should be made to establish cultures of any found. *C. carolinense* is more sensitive to changing conditions in the medium than is *A. proteus* and should be transferred to fresh medium weekly. We use Chapman-Andresen's modified Pringsheim's solution and add *Paramecium* or *Colpidium* as prey upon transfer and later as needed. A rice-grain infusion in the same saline solution can be used, so that the food species is also maintained, but cultures containing grains tend to go bad.

Further information on the cultivation of *A. proteus* and *C. carolinense* is found in the chapter by J. L. Griffin in Jeon (1973).

Pelomyxa palustris: Grows in bottom ooze with decaying vegetation, feeding on blue-greens and other algae. Does not do well with a high oxygen concentration. If it is found, it may be kept for a few months in a dish containing the ooze and decaying material in which it was found, assuming the presence of food such as filamentous blue-green algae.

Thecamoeba and *Mayorella*: All freshwater and soil species in culture except *T. proteoides* grow well on CP agar, feeding on the accompanying bacteria and small protozoa as mentioned previously. If any grow out from field material, a prey protozoan will usually be found with them.

Thecamoeba proteoides: This grows well by the method given for *Amoeba proteus*. It has also been cultivated in E+S, feeding on accompanying diatoms and other algae, which tend in time to overgrow the *Thecamoeba*.

Hyalodiscus: An amoeba which probably belongs to this seldom-reported genus has also been cultivated in E+S, feeding on accompanying diatoms. This method of culture merits further testing for medium-sized amoebae which have proved difficult to culture by other methods.

Adelphamoeba galeacystis is grown on Napolitano's agar. The amoebae feed on the accompanying bacteria, though *Klebsiella pneumoniae* may be added from time to time. The petri dish is sealed with cellophane tape.

Non-nutrient agar (NN) streaked with one of the bacterial species mentioned is the most widely applicable medium. We have found it very successful for most smaller and medium-sized amoebae (approximately 75 μm long and smaller, a few larger) except those for which other media have been mentioned previously. Both cyst-forming and non-encysting species do well on NN, but those which do not form cysts should start with more generous helpings of bacterial food to keep the cultures going. *Sappinia diploidea* and *Vannella simplex* are two larger amoebae which thrive on NN but need several heavy streaks of bacteria.

MYAS is applicable to more or less the same organisms as NN. It gives abundant growth of many non-encysting amoebae, but there is a tendency to excessive bacterial growth and reduction of encystment. Even amoebae which consume large numbers of bacteria seem to benefit from the presence of more or less bacteria-free spaces such as those found on NN but not on MYAS.

Leptomyxa reticulata (see page 143), not a member of the Gymnamoebia but a fairly common amoeboid organism, thrives and encysts on CP agar, feeding on bacteria accompanying it from the natural habitat.

Note: The field biologist or amateur who cannot easily have a large range of agar media, or finds some ingredients difficult to get, should be assured that most or all of the amoebae for which agar media are recommended will grow on the simple grass-seed agar described, though he *may* get larger numbers of some amoebae on other media. Particularly, if Cerophyl cannot be obtained, GS infusion and GS agar should be tried where the Cerophyl media are recommended, and if a bacterial culture is not easily available, where NN agar is recommended. Nor should one give up the idea of growing amoebae with aseptic precautions if an autoclave is not available, since a pressure cooker can serve the same end.

CULTURE PROCEDURES

Collected material

My usual procedure is to inoculate material collected from the field into two or three different media to permit a variety of amoebae to grow out: (1) CP or GS agar with no fluid overlay, adding approximately 1 or 2 ml of the collected water, which should soak into the agar within a few hours, along with particles of bottom material and plant matter; (2) CP or GS agar with fluid overlay, adding enough of the collected water to cover the agar to a depth of approximately 5 mm; (3) CP or GS infusion in a small dish without agar. In the last two cases, many ciliates will grow out, while the agar without overlay eliminates most ciliates and various other organisms in a few days, but permits good growth of many amoebae. If possible, it would be well to inoculate a fourth dish containing E+S to favour algivorous species.

If the collected material is leaf litter, soil, moss or other solids of non-aquatic origin, I usually distribute small quantities of it on the agar surface and add about 1 ml of sterile PJ fluid. If preferred, a soil suspension can be prepared with the same fluid and 1–2 ml added to the agar. In either case, do not cover the entire surface with the inoculum, since observation will be easier if a few clear spaces are left on to which amoebae can migrate.

Incubate the plates at room temperature in indirect light or even, except for algivorous species, in darkness. No cultures should ever be placed where direct sunlight can strike them.

After about a week the material in the first dish (agar with no overlay), should show a good number of amoebae and cysts on the agar. One may first observe the agar for presence of amoebae by turning the plate over on the microscope stage, after removing the slide-holder if possible. If the agar has been poured to the recommended depth of approximately 5 mm, it will be possible to focus on the free surface of the agar (using transmitted light) with the ×4 objective of any microscope, and the ×10 of many. One is not looking for amoebae of any particular distinctive shape; most amoebae flatten out on agar to a greater extent than in fluid. Nor can one expect to observe much, if any, protoplasmic streaming at this magnification. Contractile vacuoles can usually be distinguished even in small amoebae; this is one way to tell them from small bacterial colonies. But until one has learned by experience to recognize amoebae on an agar surface, one will have to make slides or hanging drops (see p. 22) to confirm their presence. Do not take the cover off the plate to observe the agar surface directly, unless you are looking for large amoebae and are not worried about aerial contaminants, including fungi. Using slides or hanging drops, you may by this stage already have identified some amoebae to genus.

Once amoebae have been found to be present, the initial culture should be subcultured to obtain further enrichment and leave behind any lingering ciliates and small metazoans. Rinse off the agar with 1–2 ml of fluid, using a sterile pipette, and inoculate on to a fresh surface. Alternatively, if the amoebae have migrated away from much of the inoculated material, you can cut out with a sterile scalpel a small piece of the agar surface bearing amoebae, transfer it to a fresh agar plate, and add about 0·5 ml of PJ or AS. At this stage it would be well to inoculate a plate of NN agar streaked with bacteria as well as another CP or GS plate, to obtain large numbers of some species. Do not discard the first plate inoculated with the collected material, since some amoebae, especially medium-sized and larger ones (e.g. *Thecamoeba*, *Mayorella*) may take two or three weeks to come up. You may want to subculture once more from your second culture of mixed material, but subculturing too many times before deciding which organisms if any to isolate in clonal culture may result in leaving some species behind or having them consumed by larger ones.

The procedure for mixed cultures in liquid media is similar, except that one must in those cases either make slide preparations or examine open dishes to look for amoebae.

Isolation

We are concerned here with clonal isolation, which in most cases is done by picking a single amoeba or cyst from a fresh agar surface and transferring it to an agar plate of its own. (Methods for isolating species which do not grow on agar will be described later in this section.)

Rinse the agar surface of the mixed culture with 0·5–1·0 ml of PJ or AS, using a pipette. A suspension of cysts can be streaked on to a fresh agar surface with a bacteriological loop. Active amoebae can be introduced to the isolation surface by putting a drop of suspension at each of three points at one end, then tilting the dish to form three parallel streaks. In picking out a single cyst from the surface, find one separated from the others when the streaks were made. For starting clones from active amoebae, let the plate, prepared as described, stand overnight to give the amoebae time to migrate out from the streaks and disperse. Before isolating, prepare the plates to receive the clone founders. Unless you are familiar with the medium favouring the particular amoebae, it is well to use several plates of CP agar with no added bacteria and several of NN agar with a bacterial streak. Attempt a total of at least six to ten isolations at a sitting. Find an isolated cell, focus the condenser to make a bright spot at the location of the cell, lower the stage with the coarse adjustment to give working room, make four cuts to mark off the square of agar bearing the cell, being careful not to draw in any adjacent amoebae, re-focus to check that only the one cell is indeed present on the square, and then transfer the block to the fresh plate. Add a drop or two (not more) of PJ or AS to join the block to the fresh surface. For cutting and picking out the block I use a half-spear dissecting needle, a tool adapted from the mycologists, burning it off with methylated spirit before each operation. A narrow-bladed scalpel will do, and some workers pick out a block bearing the cell with a fine pipette. One can, of course, isolate by diluting until only one amoeba is present in a sterile well or watch glass.

If you suspect that the amoeba needs not only bacteria but also accompanying protozoa (e.g. smaller amoebae) as food, include some of these on the block which you cut out, or add such protozoa or unicellular algae to the new culture. In such cases the amoeba which you are trying to isolate (e.g., *Mayorella* or *Thecamoeba*) will usually be so much larger than its intended prey that there is no chance of confusing them, and you will still have a clone of your desired species. Of course, you may end up with doubts about whether the smaller organisms present are species different from your large one or merely stages in its life cycle.

Most smaller amoebae will grow out in a week or ten days if the clone takes at all. Larger ones may take longer, and it is well to keep the plates for three weeks.

Amoebae growing in liquid media may be cloned by the dilution method mentioned above. They may also be dispersed on a fresh agar surface and picked out with a scalpel as is done with amoebae growing on agar. In this case the block of agar is put into a fresh dish of the fluid medium (with food organisms). Larger amoebae which will not grow on agar without an overlay should not be expected to survive long on such an agar surface, and if this method of isolation is used they should be picked out soon after being put on the agar.

As with mixed cultures, clones should usually be kept at room temperature (*c.* 20–23°C) in indirect light or darkness. I keep *Amoeba proteus* infusion cultures at 18–19°C. Suspected pathogens of mammals should be kept at 37°C, and this would be true from the first mixed cultures from field material, if one were seeking such pathogens. However, not all amoebae which grow at 37°C are pathogenic (Griffin, 1972), and some potential pathogens (those isolated from the eye) do not grow well at that temperature.

When subculturing from agar, lift the lid only enough to admit the pipette or scalpel. If transferring by scalpel, the desired area (preferably not more than 1 or 2 mm square) should be marked on the bottom of the dish on the microscope stage.

Most non-encysting amoebae can be maintained by monthly subculturing, though a few will require bi-weekly transfer; they should be subcultured before becoming difficult to find. Some species will survive for months if the agar does not dry out, but many will not. Most cyst-forming amoebae can be kept much longer without subculturing. For long-term storage, they should be subcultured on to agar slopes in tubes closed with a cap and kept at a temperature of approximately 5–9°C. I have obtained viable cysts of *Naegleria gruberi* from a slope more than five years old, and some non-encysting amoebae also survive for a year or more under such conditions.

HANDLING OF POTENTIAL PATHOGENS

The facultative pathogenicity of some strains of *Naegleria* and *Acanthamoeba* is mentioned in the sections on those genera. It must be emphasized that these are the two genera growing out most frequently in material collected from nature. It is not possible to avoid them in our daily life, and their existence should not, therefore, deter anyone from undertaking the cultivation and isolation of amoebae. Anyone cultivating amoebae from fresh water and soil is bound to get members of these two genera many times. It is obvious that most strains are not pathogenic, or most people are not vulnerable, or special circumstances are needed for pathological conditions to occur. Which one or which combination of these possibilities is true has not yet been finally determined. It must be said, however, that we have no assurance that facultative pathogens among free-living amoebae are limited to the genera *Naegleria* and *Acanthamoeba*. At present, pathogenicity can be determined only by animal inoculation experiments.

The aim of precautions should be to prevent any potential pathogens from invading the body, and the two known points of entry or attack are the nose and the eyes. One wishes therefore to prevent both direct transportation to nose or eyes on the fingers and production of an aerosol which might carry cysts. One should keep the head well back while rinsing with a pipette, and if there is serious suspicion that a potential pathogen is present, use of a shield or of some kind of chamber such as a fume cupboard is recommended. Glassware (e.g. pipettes, slides) should be put into a strong disinfectant solution after use and, if there is strong suspicion of pathogenicity, autoclaved. Discarded amoeba cultures should be autoclaved if possible; otherwise they should be subjected to chemical sterilization. Unfortunately, there is not yet enough information on resistance of *Naegleria* and *Acanthamoeba* to a range of chemical disinfectants to give more precise information.

OBSERVATION

Various general suggestions for observation during preliminary stages of cultivation and isolation have been given. We are concerned here with a few tips for more detailed microscopic observations of amoebae.

Although illustrations in this key include photographs made with phase-contrast and differential-interference-contrast optics, those systems are not essential for successful detailed investigation of amoebae, helpful though they may be. The essential microscopic equipment is a microscope with a good source of illumination of variable intensity and a good condenser, with a range of objectives including at least ×10, ×40 and ×90 to 100 (oil), and oculars to give a final magnification up to ×1000 or ×1250. If one can also have an objective of approximately ×63, the need for using oil on mounts of living material, with the resultant movement and other undesirable effects of pressure, will be reduced. Depending on the working distance of the ×10 objective, one may need a ×4 or ×5 objective for scanning unopened petri dishes. A green filter will be helpful at the highest power, and, if there is a choice, the condenser should be able to take a drop of immersion oil to connect it with the slide, though these last two refinements are not indispensable for identification. An ocular micrometer disc fitted into one of the eyepieces is essential. This is calibrated by means of an object micrometer slide, which will be needed only once and could perhaps be borrowed.

To observe living amoebae one should (with the exceptions mentioned below) make hanging drops. Put a drop of cell suspension on a cover glass, and place over this either a slide with a concavity or one with a ring about 2 mm deep cemented to it. Petroleum jelly is used to join the slide and cover glass, being applied to either the edges of the cover glass or the ring cemented to the slide. For species which adhere well, the drops may be turned over after two hours. For those which adhere poorly, the drop should be left overnight before being inverted. Most limax amoebae (Hartmannellidae, Vahlkampfiidae) adhere well. *Thecamoeba, Platyamoeba,* and *Mayorella* take longer and then do not adhere as well. Some species of *Vannella* adhere very poorly, and *Sappinia* seems not to attach at all. Those which adhere poorly must be observed on flat slides under cover glasses, but such preparations do not last long if made from bacterized cultures, the amoebae soon losing their normal locomotive form and rounding up, except near trapped air bubbles. The hanging drop provides oxygen and freedom from pressure. Amoebae which do not adhere to the glass may fall to the bottom of the drop and extend on the surface film as on a substratum, but of course measurements made on this curved surface will be inaccurate.

The use of hanging drops is not recommended for *Amoeba proteus, Chaos carolinense, Pelomyxa palustris* and similar large organisms. These should be examined on flat slides with the coverslip supported by petroleum jelly as described on p. 8.

For observation with phase contrast or differential interference contrast, the cover glass with its attached drop can be placed carefully and rather lightly on a flat slide, supported by its rim of petroleum jelly. The best observations are then made on amoebae attached to the underside of the cover glass.

To show cell inclusions without regard to cell shape or locomotion, whether with ordinary bright field or other optical systems, the cover glass removed from the hanging drop preparation can be pressed down more firmly on the flat slide, flattening the cells somewhat. This procedure is used for studying nuclei as well as other inclusions with bright-field optics under the oil-immersion objective. Nuclei can be found and examined in detail by a combination of slight flattening (preferably of cells already attached to the coverslip) and, if necessary, slight dimming of the illumination. Too bright light is a common hindrance to observations of nuclei. One should look for bodies resembling the nuclei in the photographs.

Floating forms are seen by putting a drop of cell suspension on a flat slide, adding a cover glass, and observing immediately. Some floating forms change to locomotive forms in a minute or two; in other species the floating form may persist for hours.

Cysts are examined in ordinary wet mount preparations. Especially in this case, phase contrast is a hindrance rather than a help, and observations with ordinary bright-field optics are much more satisfactory.

Staining techniques are not within the province of this key, and the reader is referred to books on that subject in the bibliography (Galigher & Kozloff, 1971; Grimstone & Skaer, 1972) and to Page (1967). We prefer to fix the amoebae on cover glasses and stain them in coverglass staining jars (Columbia jars). DPX mountant (BDH Chemicals Ltd) has proved the most satisfactory of those which we have used.

Techniques for inducing flagellate transformation in certain vahlkampfiid amoebae are given in the introduction to the family Vahlkampfiidae, since they are meaningful only in connection with the discussion of genera in that family.

USING THE KEY

The taxonomy of amoebae and the criteria upon which to base it have been the subject of much controversy. The system employed here is based on a synthesis of criteria, but the criteria themselves will be explained only from a practical viewpoint. Those wishing a further discussion of principles are referred to Schaeffer (1926), Singh (1952, 1975), Jahn & Bovee (1965), Bovee & Jahn (1965, 1966), Pussard (1973b), and Page (1967, 1968, 1972, 1974, 1976). Future keys will undoubtedly include new characters not yet investigated widely enough for inclusion here, though such characters may, unless correlated with more easily observed ones, make difficulty for the worker with only simple equipment: ultrastructure of scales in *Cochliopodium* and *Mayorella*, antigenic comparisons and DNA base compositions in *Acanthamoeba* and *Naegleria*.

The following explanation of criteria deals with their actual use in this key.

Locomotive form and behaviour. These cannot be separated, since the manner of locomotion produces the appearance, while the underlying structure affects the manner of locomotion. For practical purposes the terms *hyaloplasm* and *granuloplasm* are employed for the major observed divisions of the cytoplasm, rather than ectoplasm and endoplasm, plasmagel and plasmasol. *Eruptive activity* is that which can be seen as a burst of cytoplasmic movement rather than a more steady flow, though the latter may result in some bulging or even the production of long pseudopodia. *Rolling locomotion*, as in the family Thecamoebidae, involves a tractor-like movement of the cell surface (especially noticeable when there are adherent particles) and usually only gradual changes of form, e.g. expansion in a new direction or a pulling up of the posterior end, broadening the amoeba. Such rolling movement of the cell surface is not always obvious on a cursory examination, and the actual shape and any surface folds are more immediately useful criteria for identification.

Amoebae which move as one mass, without several distinct pseudopodia, are *monopodial* (Figs 2, 8A, 13A, 20A); those with more than one advancing point are *polypodial* (Figs 5D, 6A, 7A). As a practical matter, the terms monopodial and polypodial are used here only to distinguish among amoebae of a tubular (cylindrical or sub-cylindrical) form. *Limax* amoebae (e.g. Figs 13A, 20A) are monopodial cylindrical amoebae, and the term does not apply to noticeably flattened amoebae. This term, as well as all others describing over-all form, is meant to describe amoebae in active locomotion, that is, actually progressing in a given direction. Many limax amoebae put out a lateral pseudopodium to change direction and thus temporarily have more than one pseudopodium, but their characteristic form is monopodial. (*Clavate* is another term sometimes used to describe elongated monopodial forms.)

Many amoebae produce several or many narrow (conical or filose) projections from a broad hyaline zone which may itself be considered one large pseudopodium. To circumvent difficulties in terminology, it is sometimes said that such amoebae produce fine or filose pseudopodia from a lobopodium. However, when there is any chance of ambiguity this work employs the term *sub-pseudopodia* for such narrow projections. The production of such sub-pseudopodia is not considered to make an amoeba polypodial, but the monopodial/polypodial dichotomy is at any rate not used for such amoebae. Sub-pseudopodia range from fairly lobose to very fine. All are technically conical, though that word is not usually applied to distinctly filose ones, and all are hyaline for all or most of their length. (We are, again, discussing only the locomotive form and only the subclass Gymnamoebia.) *Mamilliform* sub-pseudopodia (Figs 54A, C, 57A) have a fairly broad pyramidal base and a smaller conical tip. Sub-pseudopodia found in many *Mayorella*, which have a distinctly rounded tip and do not taper strongly from their base but have a fairly uniform diameter for some length, are described as *digitiform* (Figs 50C, 51D). There are sub-pseudopodia (e.g., in *Acanthamoeba* and *Dinamoeba*) which taper more strongly than the digitiform ones and are also more likely to bend; these have been called *acanthopodia* (Figs 44, 59A). They intergrade with finer pseudopodia until the unquestionably *filose* ones (e.g., *Filamoeba*, Fig. 42A–C) are reached. Very short finely tipped projections which may have a broader base are called *micro-spines* (Fig. 43A, H) in this key. Acanthopodia, filopodia, and those between them may be *furcate*, that is, branched or clustered two or three on a single narrow base.

Lengths (L) and length:breadth ratios (L/B) are taken from

measurements of actively moving forms. *Length* is the anterior-posterior direction, anterior being understood as the end leading in locomotion at the moment of measuring. While cell volume can be calculated more accurately from rounded cells, experience shows these lengths to be of considerable practical value and as constant as can be expected with a changing shape. Poorly nourished cells are smaller, as are those which have just divided, and multinucleate cells are considerably larger. When a size is given without further qualification (e.g. 30–50 μm), it is the length in normal continuing locomotion. When the length in locomotion was not given by the original author, and the present author has not investigated the species at first hand, whatever measurement the original author reported is used. Sometimes this was simply '*size*' without definition of that term; in such cases 'size' is used in inverted commas. Sometimes the '*diameter*' given appears to be the length of a locomotive form rather than the diameter of a rounded form. All measurements should be considered approximate, with the range subject to expansion.

The posterior end of an amoeba is called the *uroid*, particularly if it is anything but a simple, smooth, rounded end. If it has a knobby appearance, often with little rounded bumps, it is a *morulate* uroid (Figs 3A, 53, 54B). It may have a few trailing *uroidal filaments* (Fig. 8D) produced by adhesion to the substratum. A cluster of a large number of such filaments, varying somewhat in length and noticeably flexible, whether or not on a distinct bulb, is an *adhesion uroid* (Fig. 13A). A bulb-like posterior end (often set off slightly from the main mass by a 'neck'), bearing rather short, fine but fairly rigid projections, is a *villous-bulb* uroid (Figs 2A, 21D, E).

In some parts of the key the term *dorsal* is used for the free surface of the amoeba, that is, the one opposite to the surface applied to the substratum.

Floating forms, produced when an amoeba is disturbed and best observed when a drop of cell suspension is first put on a slide, are characteristic of the different groups but only in a few cases of use in identification. *Radiate* pseudopodia, usually hyaline for most of their length, project from a central mass, sometimes symmetrically, often asymmetrically. Those of *Vannella* (Fig. 35D) taper to a rather sharp point (though they may sometimes be blunt at first and then extend to the fine form), while those of *Platyamoeba* (Fig. 38D, I), when present at all, are quite blunt and less tapering.

Mayorella (Fig. 51G) often has a floating form resembling that of *Vannella*, but they will not be confused when the locomotive form is seen. Some species of *Vannella* remain in the floating form a long time, perhaps hours, and it was undoubtedly such persistent floating forms which gave rise to the old generic name *Astramoeba*. Limax amoebae usually have only irregularly rounded and very transient floating forms. Members of the family Amoebidae (including in this key the limax genus *Trichamoeba*) have floating forms with radiating pseudopodia, granular for most of their lengths, with hemispherical tips (Fig. 6E), much resembling a narrow locomotive pseudopodium of *Amoeba proteus*.

The *amoeba-to-flagellate transformation* as a taxonomic criterion is discussed under the family Vahlkampfiidae.

Nuclear structure and nuclear division patterns have long been used as taxonomic criteria and as such have been the centre of controversy for more than half a century. Nuclear structure is used in this key whenever necessary. The most common kind of nucleus in amoebae is the *vesicular* nucleus (Fig. 1A), with a single central nucleolus (endosome, karyosome) surrounded by what appears to be a clear space, though in larger nuclei of this kind the layer of chromosomal material can sometimes be seen faintly *in vivo* between the nucleolus and the nuclear membrane. Some vesicular nuclei have two or three nucleolar bodies arranged peripherally, that is, against the nuclear membrane; these are designated in this key as *multiple peripheral nucleoli* (Fig. 39). In the families Amoebidae (except in a few questionable cases) and Pelomyxidae, the apparent nucleolar material consists of many small particles or granules, often distinctly seen to be arranged in a layer just inside the nuclear membrane; these nuclei are designated in the key as *granular* (Figs 3B, 6D, 8B). All species whose nuclei are not described have a vesicular nucleus with a single central nucleolus; all are uninucleate unless described otherwise.

It is hoped that the user will seldom have to undertake stained preparations for mitotic figures, although hundreds of such preparations have gone into the elucidation of relationships set forth here. The principal case where such preparations may be necessary is the identification of limax amoebae, if it cannot be decided whether the isolate is a vahlkampfiid or a hartmannellid on the basis of the eruptive/non-eruptive locomotive dichotomy. Some mitotic

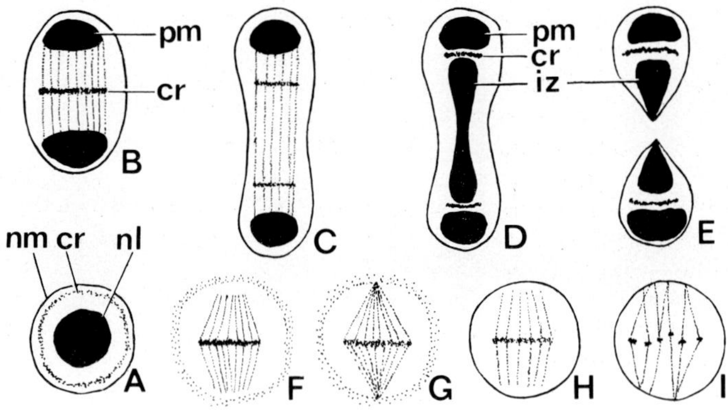

Fig. 1. Nuclear division figures. A, vesicular nucleus with central nucleolus. B, metaphase of promitosis. C, anaphase of promitosis without interzonal bodies. D and E, late stages of promitosis with interzonal bodies. F and G, metaphase of mesomitosis without nuclear membrane present. H and I, metaphase of mesomitosis with nuclear membrane present; I, with poles apparently dispersed, probably an earlier stage than H. *cr*, chromosomal material; *iz*, interzonal body; *nl*, nucleolus; *nm*, nuclear membrane; *pm*, polar mass.

patterns are illustrated in Fig. 1. Terms used to describe mitotic patterns are (1) *promitosis*, in which the nucleolus elongates and divides into two polar masses and the nuclear membrane remains intact; (2) *mesomitosis*, in which the nucleolus disintegrates and the nuclear membrane also does sooner or later, giving a mitotic figure like that in metazoans except for the absence of centrioles; (3) *metamitosis*, a pattern like that of metazoans, with centrioles or centriole-like structures. Pussard (1973b) has proposed changes in these definitions.

Centrioles were reported from amoebae early in this century, but subsequent investigations, including ultrastructural studies, did not bear this out as far as classical centrioles are concerned. Therefore, one would not expect to find nuclear division in any amoebae designated as metamitotic if the classical ultrastructure for a centriole is required. Pussard (1965, 1966, 1973b) has reported *centrospheres* in *Acanthamoeba*, *Cochliopodium*, and *Gocevia* (Fig. 26E), these bodies apparently functioning as polar centres, and an ultrastructural study supports his claim for at least one species of *Acanthamoeba*. However, the centrosphere is not used as a character for identification here.

Cysts. The value of cysts as a taxonomic criterion is abundantly illustrated in this key. Apart from their structure, one would expect cyst diameter to be a reliable characteristic, since an individual cyst does not change shape, though one must allow for occasional cysts of very atypical shape. In many cyst walls, distinct outer (*ectocyst*) and inner (*endocyst*) layers are present. Cells may leave the cyst by bursting or apparently digesting the wall, but *Naegleria gruberi* leaves by a pre-formed pore from which a plug is removed, and *Acanthamoeba* removes an *operculum* (lid-like cap) from the exit hole and takes it into the cyst, where it can later be seen.

Various *cytoplasmic inclusions* other than centrioles have been described and employed as taxonomic characters. The DNA-containing *Nebenkörper* is found only in the marine genus *Paramoeba*. Crystals have been used as a criterion, though all appear to be the bipyramids and plates identified in some amoebae as triuret, an apparent waste product. The occurrence of 'sand grains' in *Pelomyxa palustris* is easily observed, but since such mineral particles may be absent from some individuals, their presence or absence in distinguishing species must be used with care. *P. palustris* lacks mitochondria; presumably other species of *Pelomyxa* also do (if indeed there are any others), but at any rate this character would be difficult to determine in observing *in vivo* amoebae whose cytoplasm is so packed with other inclusions. The occurrence of zoochlorellae is reliably known in only one species of naked amoeba, *Mayorella viridis*.

Very often a single criterion is insufficient to identify an amoeba, though described species of some genera, such as *Thecamoeba*, should be easily identified by observation and measurement of a moving cell. However, different genera, and certainly different species of the same genus, may have similar floating forms. Distinctive cysts may immediately identify a few genera and species, but many different organisms (not just amoebae) have small, smooth-walled, spherical cysts. One must often, therefore, observe various characters and stages in order to identify an amoeba to genus or species. Because in some cases one needs all possible information, the keys contain much information not needed for a simple dichotomy but useful or even essential as confirming evidence, e.g. nuclear size. This additional information is also intended to prevent misidentifications of similar but undescribed species.

With a few amoebae (*Thecamoeba* is again an example), it may be possible to make a specific identification from a single healthy individual. With most, however, one should observe several individuals under conditions which do not distort their normal form and behaviour (see p. 22). There is variation also among cysts of a species. One would certainly want to see a good number of cysts before deciding whether an isolate is *Acanthamoeba castellanii* or *A. polyphaga*. Fortunately, if one has any cysts of these at all, one is likely to have many. Otherwise one can subculture and obtain enough.

In the notes of reported geographic distribution, 'Europe' includes Britain, though not all species noted for Europe have been reported for Britain; the mention of 'Europe' indicates reports of occurrence from more than one country.

Recognizing that this key will not necessarily make the identification of most amoebae easy, I hope that it will make the identification of many amoebae possible.

ABOUT THE ILLUSTRATIONS

Photomicrographs have the disadvantage of showing only what is in focus at one time and is optically dense to the system used, but the advantage of giving a view relatively unobscured by the author's interpretation. Drawings have the advantage of not being limited by depth of focus or even by the observations that can be made in a moment on a single individual. This key uses drawings where photographs were unavailable or obviously inadequate. Some photographs were made with bright-field optics, some with phase contrast, some with differential interference contrast. The bright-field photos are meant to show amoebae as most users will see them. However, even if most workers are not equipped with interference contrast, photos made with that optical system are often more informative than others, and they give a much more accurate picture than drawings. After all, no one sees a drawing under the microscope with any optical system.

It must be emphasized that the illustrations cannot show all possible variations of the locomotive shape of a species. They are meant to be used in conjunction with the text. **The user should not expect that every amoeba of a described species will have a form fitting into the illustrated one.** Where it seemed necessary and was possible, multiple illustrations of a species have been used.

Magnifications of the photographs are given in the captions. The magnification of each drawing or set of drawings is shown by a line with a figure giving its length in micrometres (μm).

KEY TO FAMILIES

1 Locomotive form a branched or unbranched cylinder, not noticeably flattened and not producing sub-pseudopodia during normal locomotion— **2**

— Locomotive form more or less flattened, discoid, or globose, but not an unadorned cylinder; often with sub-pseudopodia— **6**

2 Usually polypodial, with pseudopodial tips hemispherical and usually with hyaline cap; nucleus granular; pseudopodia of radiate floating form granular for most of length, not tapering strongly (Fig. 6E); if usually monopodial in locomotion, then commonly with villous-bulb uroid, but with nucleus, locomotive pseudopodia, and floating form characteristic of family— AMOEBIDAE, p. 35

— Limax amoebae without characters of above family— 3

3 Multinucleate, occasionally binucleate, rarely uninucleate, large or medium-sized amoebae with thickly limax-like or elongately ovoidal form; often true bidirectional streaming of cytoplasm; hemispherical anterior end usually without hyaline cap; often a villous uroid; symbiotic bacteria, usually mineral particles in cytoplasm— PELOMYXIDAE, p. 46

— Normally uninucleate; no rearward streaming of cytoplasm— 4

4 Endosome of nucleus a small granule or cluster of granules; mature cyst (if produced) with 4 or 8 nuclei; all parasitic except one species, and that anaerobic— ENTAMOEBIDAE, p. 50

— Diameter of endosome usually half that of nucleus or greater; cysts, if any, normally uninucleate, or, in one species, very often binucleate; known species aerobic— **5**

5 Locomotion by a succession of more or less discrete, eruptive, hyaline, hemispherical waves, often to alternate sides at anterior end, with hyaloplasm sometimes spilling rearward along side; flagellate stage in most genera; nuclear division promitotic— VAHLKAMPFIIDAE, p. 51

— Locomotion by generally steady flow, sometimes with gentle hemispherical bulging to either side at anterior end; no flagellate stages; nuclear division mesomitotic— HARTMANNELLIDAE, p. 68

6 (1) With distinct, flexible covering, often doubly contoured in profile (Fig. 26A) or with perceptible punctation (Fig. 25A–C), over hump-like or globose granular mass; flattened hyaline border which may completely surround granular mass or, if anterior only, may produce short conical or pyramidal projections; general shape discoid (in common forms) or globose; several to many prominent bipyramidal crystals in common species; locomotion not involving rolling movement of covering— COCHLIOPODIIDAE*, p. 76

— Without such a covering; if with pellicle-like layer, then having extensive anterior hyaline zone with more or less smooth edge and rolling movement of surface— **7**

7 Flattened, with no sub-pseudopodia produced during normal locomotion and without flattened border entirely surrounding granular hump— **8**

— Few or many coarse or fine sub-pseudopodia produced during normal locomotion, or sometimes lacking fine pseudopodia but with regularly discoid shape and flattened hyaline border often surrounding granular hump— **9**

* Although included in this key, the Cochliopodiidae are not classified in the Gymnamoebia.

8 No obvious pellicle-like layer; irregular and rapidly-changing form, sometimes spatulate or flabellate, and often with eruptive cytoplasmic flow— FLABELLULIDAE, p. 85

— Outline more or less symmetrical and regular, often oblong, ovate, flabellate, with anterior edge usually more or less smooth in most active locomotion; pellicle-like layer discernible in many; rolling movement of surface in locomotion— THECAMOEBIDAE, p. 86

9 (7) Regularly discoid, flattened ovoid, or fan-shaped, usually with breadth the greatest dimension, with central, post-central, or posterior granular hump surrounded completely or anteriorly and laterally by flattened hyaline border, sometimes with slender, short, conical pseudopodia; often apparent rolling movement of entire cell; no tectum of scales— HYALODISCIDAE, p. 105

— Not a regular, broad, discoid form— **10**

10 Several to many distinctly filose sub-pseudopodia or microspines produced from a distinct hyaline zone; known species cyst-forming but not with endocyst polygonal or stellate, and not excysting by removal of operculum— ECHINAMOEBIDAE, p. 107

— Sub-pseudopodia conical, either digitiform or sharply tapering to narrow tip or pyramidal, but not consistently filose— **11**

11 Sub-pseudopodia usually tapering to a fairly fine tip, flexible, sometimes furcate; cyst with inner wall often polygonal or stellate, outer wall often rippled or wrinkled; excystment with removal of operculum (see page 29)— ACANTHAMOEBIDAE, p. 112

— Sub-pseudopodia digitiform cones, mamilliform pyramids, or elongated rods, markedly tapering and furcate in only one genus; cysts uncommon, if present not as described for previous family— PARAMOEBIDAE, p. 118

KEYS TO GENERA AND SPECIES

Family AMOEBIDAE Diesing, 1848

1 Almost always monopodial (limax-like), commonly with villous-bulb uroid; uninucleate— TRICHAMOEBA

— Commonly, though not invariably, polypodial; villous-bulb uroid not reported— 2

2 Multinucleate— CHAOS

— Uninucleate— 3

3 One pseudopodium dominant at any one time, with somewhat shorter pseudopodia originating from main stem— AMOEBA

— Several more or less equal pseudopodia, with no one dominant, produced anteriorly and antero-laterally from common posteriorly-situated mass; pseudopodia sometimes flattening and spreading, especially toward tips— POLYCHAOS

Genus TRICHAMOEBA Fromentel, 1874

Usually limax-like, commonly with villous-bulb uroid; locomotion by steady flow with occasional slight antero-lateral bulging; floating form, where known, with long, blunt, mostly granular pseudopodia; nucleus granular; normally uninucleate.

1 Usually more than 200 μm— 2

— Usually 150 μm or less— 3

2 Up to 500 μm, often smaller, with L/B often 3; nucleus to about 36 μm (Fig. 2A)— **Trichamoeba villosa** (Wallich, 1863)

Fresh water; England.

— 200–300 μm, generally thicker than above, with L/B 2·5–3; nucleus 40–60 μm; endoplasm with few granules (Fig. 2B)— **T. myakka** Bovee, 1972

Fresh water; U.S.A.

3 (1) Length 100–150 μm; relatively slender, with L/B 2–5; nucleus 13–17 μm; numerous crystals up to 4·5 μm long (Fig. 2C)— **T. osseosaccus** Schaeffer, 1926

Fresh water; U.S.A.

— Length 80–150 μm; broader than above, with L/B 2·25–3; nucleus 8–15 μm; crystals shorter, more plate-like than above (Fig. 2D)— **T. cloaca** Bovee, 1972

Sewage; U.S.A.

Fig. 2. *Trichamoeba*. A, *T. villosa* (after Wallich). B, *T. myakka*. C, *T. osseosaccus*. D, *T. cloaca*. (B–D after Bovee.) *n*, nucleus.

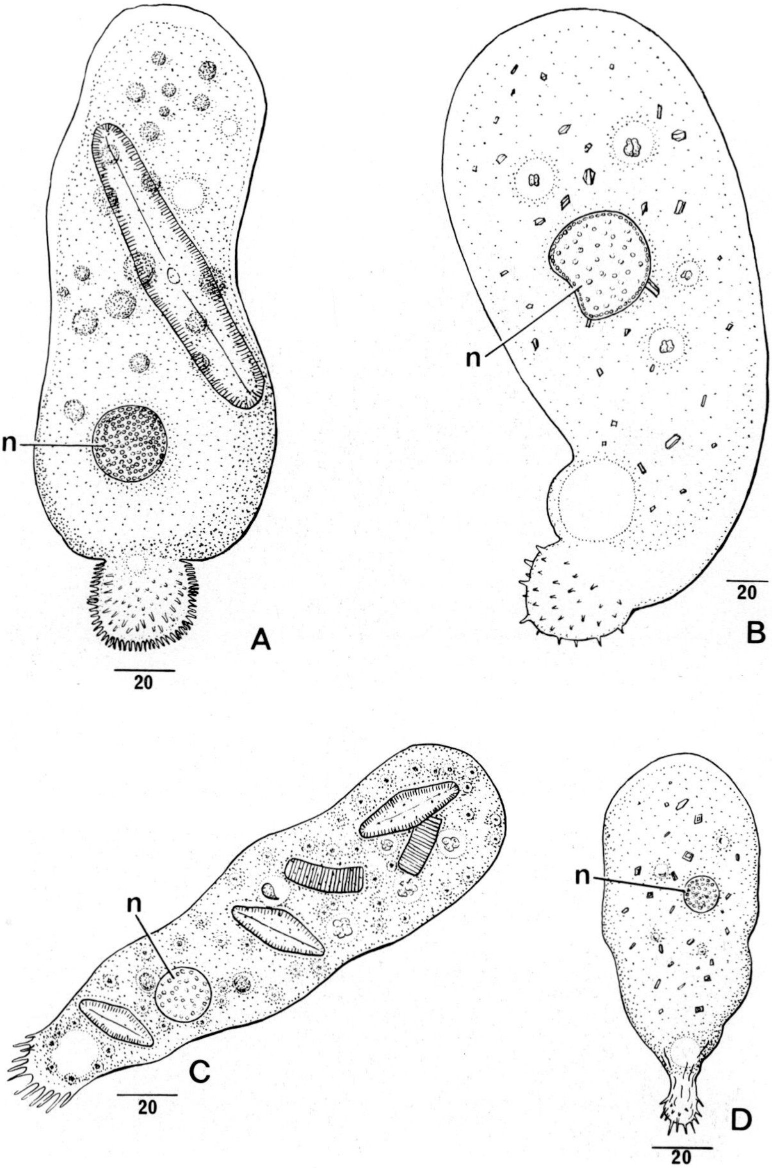
n
20
A
n
20
B
n
20
C
n
20
D

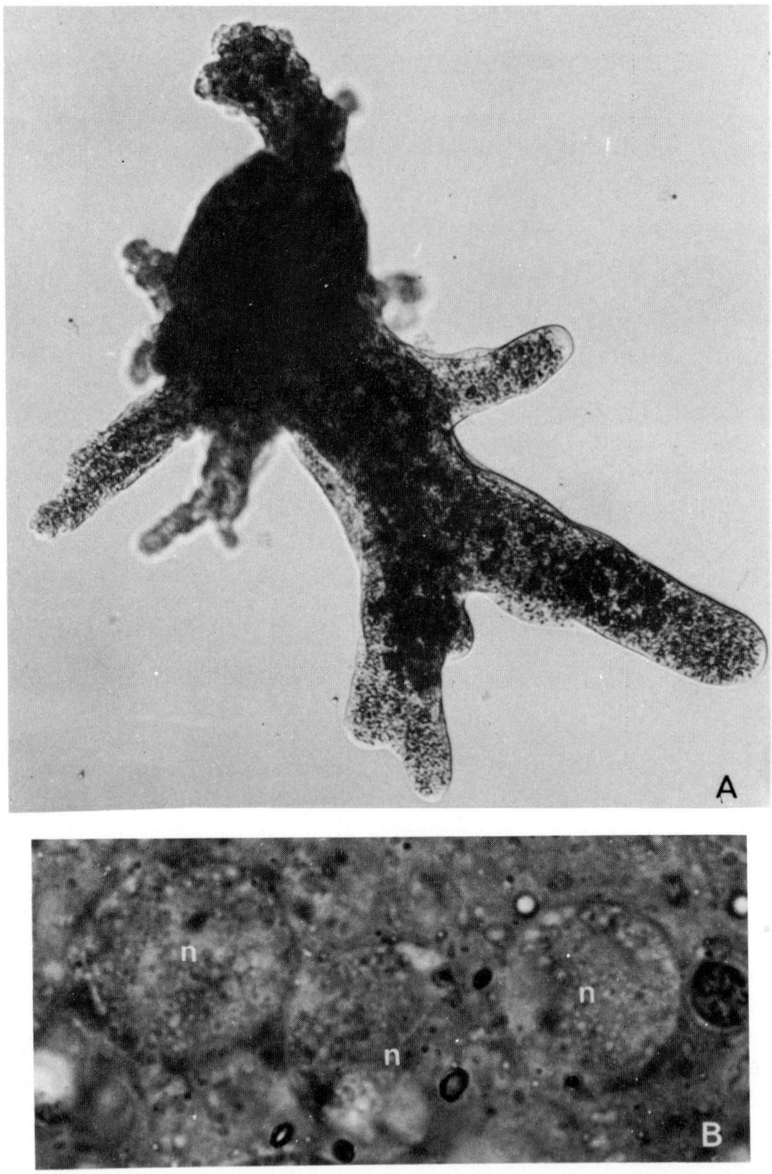

Fig. 3. *Chaos carolinense*. A, locomotive form (×100). B, nuclei (*n*) and cytoplasmic inclusions (×1,000).

Genus CHAOS Linnaeus, 1767

Large, multinucleate amoebae (hundreds of nuclei) resembling the genus *Amoeba*. Commonly polypodial but often monopodial; hyaline cap formed, as tubular pseudopodium elongates, but often less prominent than in *A. proteus*. Not common.

Penard (1902) described two large multinucleate amoebae which may belong to this genus but need further investigation: *Amoeba laureata*, which rarely took on a polypodial form similar to *A. proteus*, and *A. nobilis*, which had the pseudopodial character of *Polychaos*.

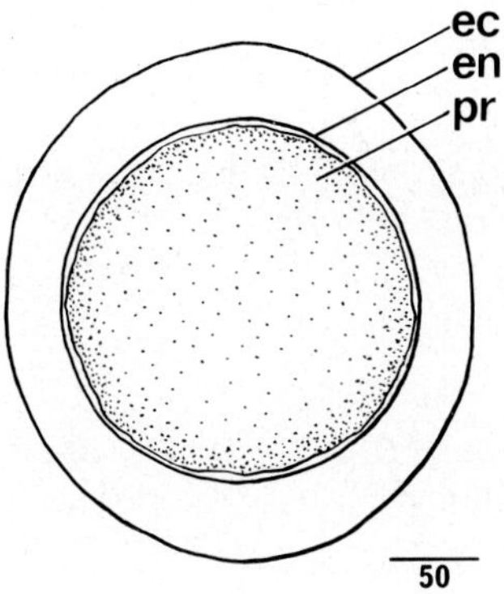

Fig. 4. *Chaos illinoisense*, cyst (after a photograph by Kudo). *ec*, ectocyst; *en*, endocyst; *pr*, protoplast.

1 Polypodial forms *c.* 700 μm to 2 mm; monopodial forms up to 3 mm or rarely more; nuclei ellipsoidal, granular, with greater diameter *c.* 20–30 μm; cyst reported (Fig. 3)— **Chaos carolinense** (Wilson, 1900)
Fresh water; U.S.A.

— Smaller, *c.* 500–1000 μm; monopodial forms up to 1·5 mm; diameter of nuclei *c.* 14–16 μm; cysts smooth, spherical, ovoid or ellipsoid, with ectocyst and endocyst widely separated, diameter 250–350 μm (Fig. 4)— **C. illinoisense** (Kudo, 1950)
Fresh water; U.S.A.

Genus AMOEBA Bory de St. Vincent, 1822

Commonly polypodial, with several tubular pseudopodia with hemispherical tips and usually hyaline caps, one pseudopodium dominant at any one time; may become monopodial in rapid locomotion; nucleus granular in most described species.

An amoeba which on brief observation might be erroneously identified as belonging to this genus is *Thecamoeba proteoides*, q.v.

1 Polypodial or monopodial forms 15–20 μm long; nucleus discoid, of obscure structure, 3–4 μm (Fig. 5A–C)—
Amoeba diminutiva Bovee, 1972

— Much larger— **2**

2 Length 500–1200 μm; nucleus spherical or elongated; no crystals— **A. navina** Shah, 1971

So far reported only from India; 'the largest mononucleate form among Amoebidae known so far.' (Shah, 1971)

— Somewhat smaller, with discoid granular nucleus— **3**

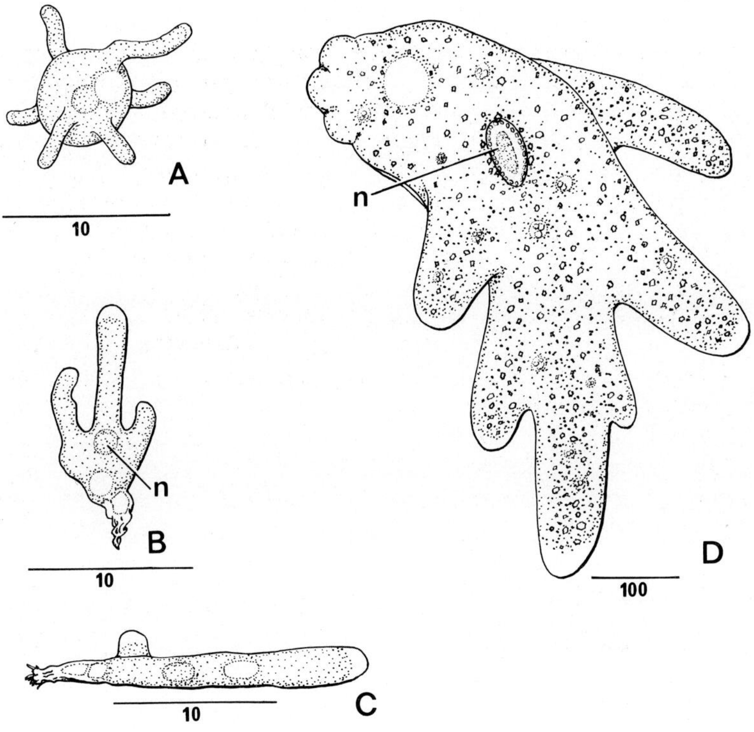

Fig. 5. *Amoeba*. A–C, *A. diminutiva* (A, floating) (after Bovee). D, *A. kerri* (after Taylor). *n*, nucleus.

3 Pseudopodia with conspicuous longitudinal ridges; often a morulate uroid; laboratory strains 220–750 μm, with means slightly under 500 μm; nucleus with greatest diameter 34–48 μm; many crystals (Fig. 6)— **A. proteus** Leidy, 1878

Fresh water; North America, Europe. Leidy's (1879) description is the first of an organism recognizable as that known by this name today. *A. lescherae* Taylor & Hayes, 1944, and *A. discoides* Schaeffer, 1916, are almost certainly synonyms of *A. proteus*.

— No ridges; maximum length 576 μm (Fig. 5D)— **A. kerri** Taylor, 1947

Fresh water; Scotland. Taylor (1947) reported a complex and questionable life cycle.

Fig. 6. *Amoeba proteus*. A, polypodial locomotive form ($\times$200). B, more or less monopodial locomotive form ($\times$200). C, locomotive form showing surface ridges ($\times$100). D, nucleus ($\times$1,000). E, floating form ($\times$100).

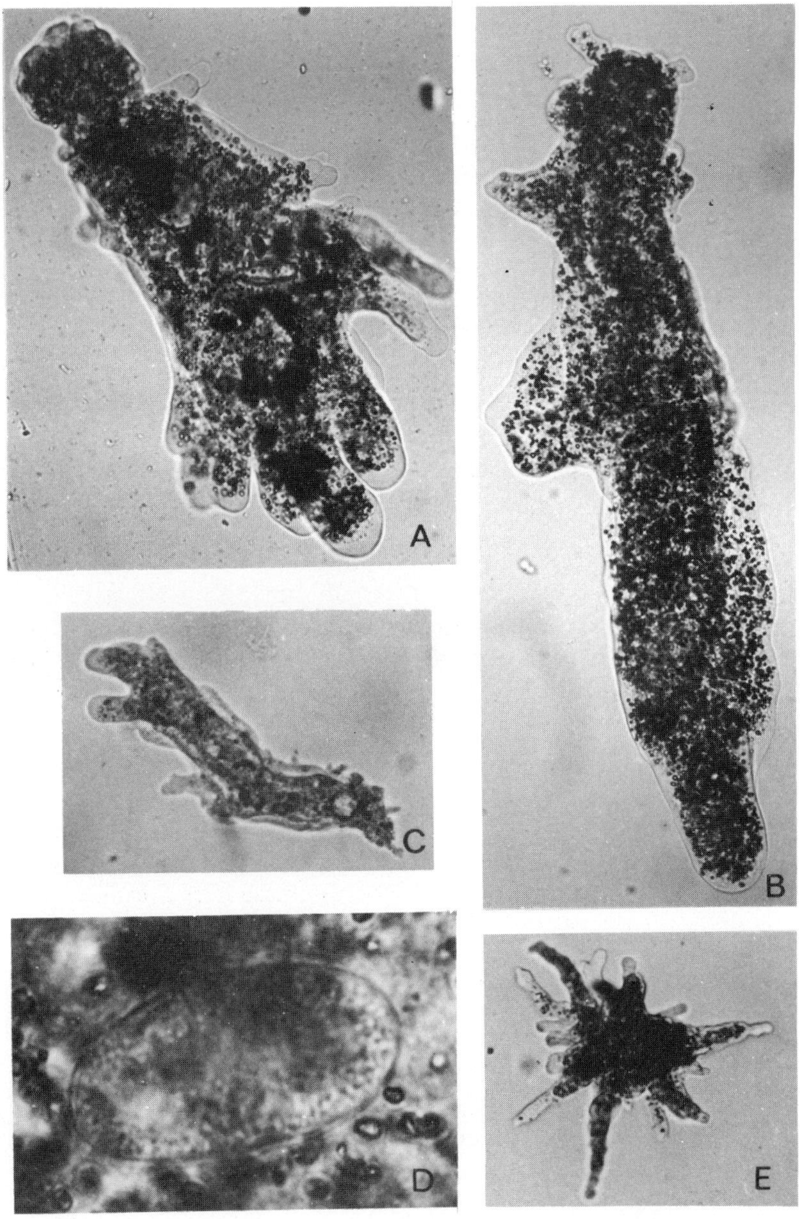
A
B
C
D
E

Genus Polychaos Schaeffer, 1926

Usually polypodial, with several tubular pseudopodia, more or less equal and with no one dominant, proceeding anteriorly and antero-laterally from common posterior mass; pseudopodia sometimes flattening and broadening towards ends; may be monopodial in rapid locomotion; no dorsal ridges; nucleus granular or with other arrangements of nucleolar material reported.

1 Nucleus granular— **2**

— Other nuclear structure— **3**

2 Palmate form *c.* 200–400 μm; rarely monopodial; nucleus typically a prolate spheroid, 30–40 μm (Fig. 7A)— **Polychaos dubium** (Schaeffer, 1916)
Fresh water; U.S.A.

— Palmate form 220–450 μm, with pseudopodia tending to be shorter and less numerous than those of *P. dubium*; monopodial form 250–550 μm; nucleus shallowly to deeply cup-shaped, 27–35 μm. (Fig. 7B, C)— **P. nitidubium** Bovee, 1970
Fresh water; U.S.A.

3 (1) Palmate form averaging 140 μm; usually several blunt, hyaline caudal appendages; no monopodial form or distinct crystals reported; nucleus, averaging 18–19 μm, with nucleolar material divided into two equal crescents on opposite sides of nucleus (Fig. 7D)— **P. fasciculatum** (Penard, 1902)
Fresh water; Switzerland.

— Palmate form 45–60 μm; monopodial form 80–100 μm; usually a uroid of clear spines; floating form with several pointed, granular, radiate pseudopodia; nucleus 6·5–7 μm with central endosome apparently a cluster of particles (Fig. 7E, F)— **P. timidum** Bovee, 1972
Fresh water; U.S.A.

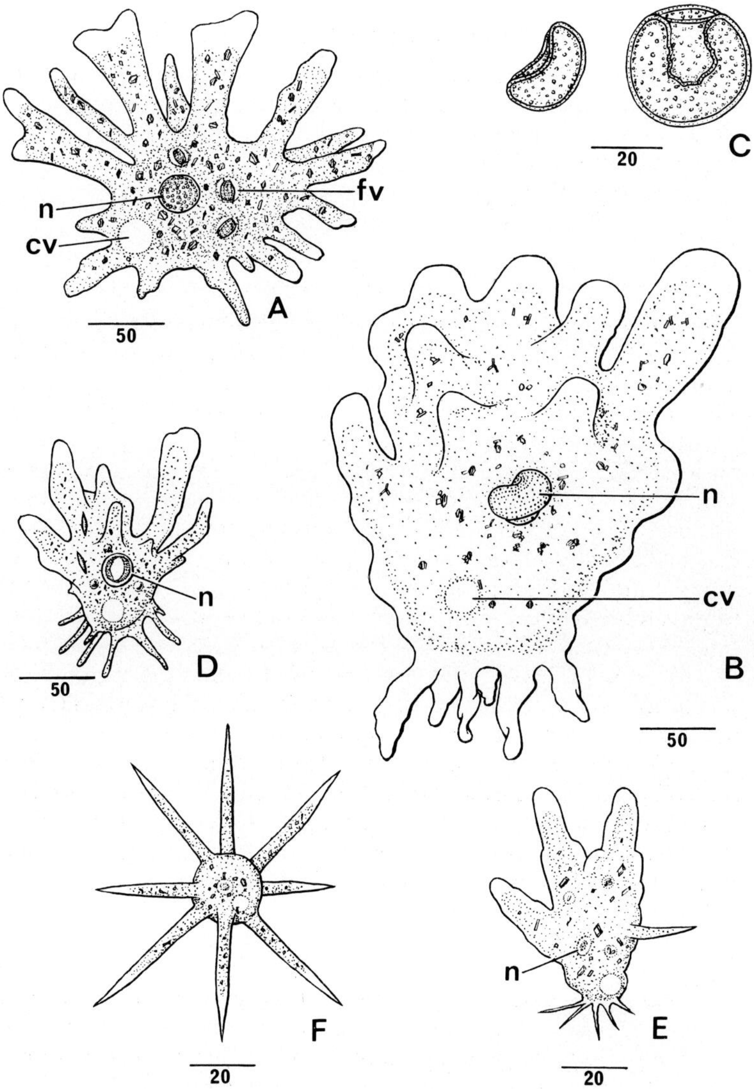

Fig. 7. *Polychaos*. A, *P. dubium*. B, C, *P. nitidubium* (C, nuclei) (after Bovee). D, *P. fasciculatum* (after Penard). E, F, *P. timidum* (F, floating) (after Bovee). *cv*, contractile vacuole; *fv*, food vacuole; *n*, nucleus.

Family PELOMYXIDAE Schulze, 1877

Usually large or very large, thickly limax-like and usually without a hyaline cap, usually multinucleate, and with other unusual characteristics described for the one well-established genus and one well-studied species.

Genus Pelomyxa Greeff, 1874

Thickly limax-like or ovoid, usually large, moving rather slowly by fountain streaming or hemispherical eruption, commonly without hyaline cap except occasionally at start of locomotion, and frequently with uroid of short villi; may put out short, conical, hyaline pseudopodia from much of surface, usually when not in locomotion; usually multinucleate, with most common type of nucleus granular but variation in structure reported; no mitochondria in one well-studied species; symbiotic bacteria in cytoplasm; feeding principally on non-motile plant matter, mostly algae, and with algal remnants and mineral grains often prominent cytoplasmic constituents; living in polysaprobic habitats such as bottom mud of ponds and ditches.

The question of species is very unsettled, although more than 20 named species have been attributed to the genus. Collected specimens vary greatly in size, cytoplasmic inclusions, and even nuclear structure, but recent workers have tended to consider all their specimens as *P. palustris*, even the few uninucleate ones found. The following key includes several named species besides *P. palustris*; those included seem the ones most likely to be separate species, if any are. The question will not be settled until a reliable culture method is devised and reproduction is satisfactorily described.

1 Usually hundreds of nuclei; usually mineral grains; no contractile vacuole; length from somewhat less than 100 μm to 2 mm or more (usually more than 150 μm); nucleus approximately 8–18 μm, commonly 12 or 13 μm (Fig. 8A–C)—
Pelomyxa palustris Greeff, 1874

Fresh water; Europe, North America.

— One or few nuclei; generally smaller than above— **2**

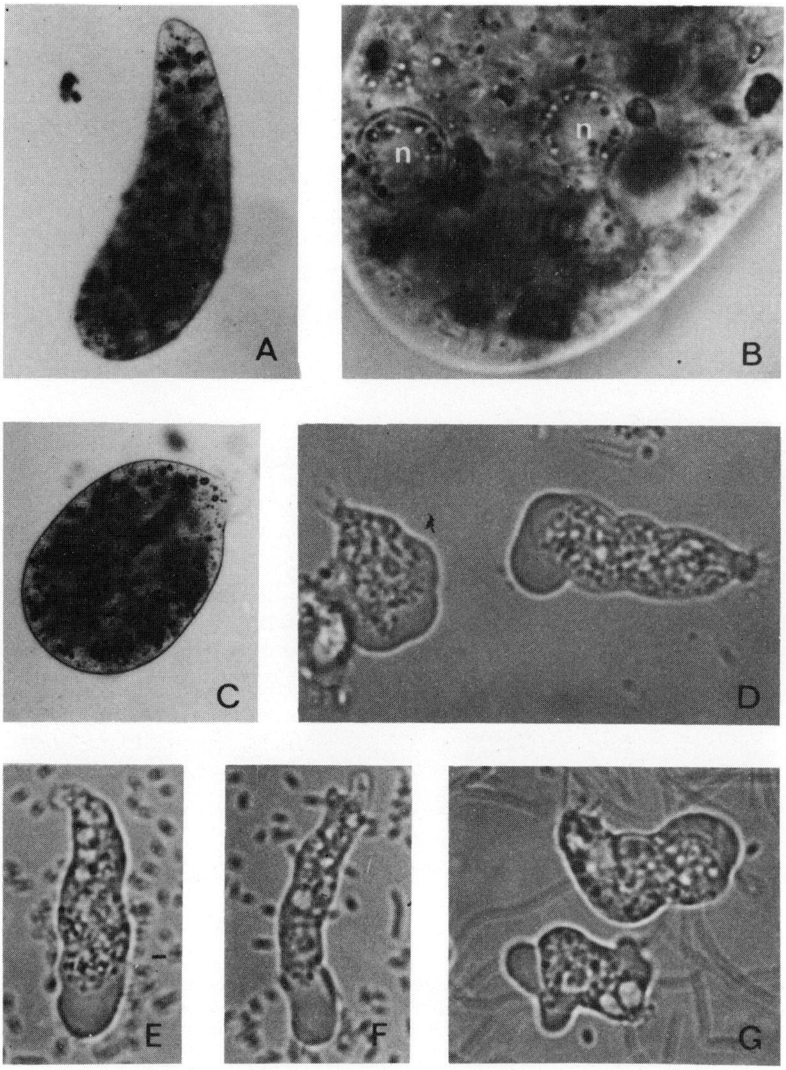

Fig. 8. A–C, *Pelomyxa palustris*, rather small forms. D–G, locomotive forms of *Vahlkampfia*. D, *V. avara*. E, F, *V. aberdonica*. G, *V. enterica*. (A, C, ×200; B, D–G, ×1,000.) *n*, nucleus.

2 Nuclei 1–12, diameter *c.* 25 μm; no mineral grains but containing fragments of decayed leaves; contractile vacuoles; often with short, irregularly lobed surface projections; oval or rounded, length usually below 400 μm (Fig. 9A, B)—
Pelomyxa belevskii Penard, 1893

Fresh water; Switzerland.

— Usually binucleate— 3

3 Binucleate, sometimes uninucleate; no mineral grains reported; often many 'starch grains'; numerous small contractile vacuoles; length approximately 75 μm; nuclear diameter *c.* 7 μm (Fig. 9C)— **P. schiedti** Schaeffer, 1918

Fresh water; U.S.A.

— Binucleate, rarely uninucleate; often some mineral grains; 1–3 contractile vacuoles; distinct hyaline border; length approximately 200 μm, reaching 300 μm; nuclear diameter 30–33 μm (Fig. 9D, E)— **P. binucleata** (Gruber, 1885)

Fresh water; Germany.

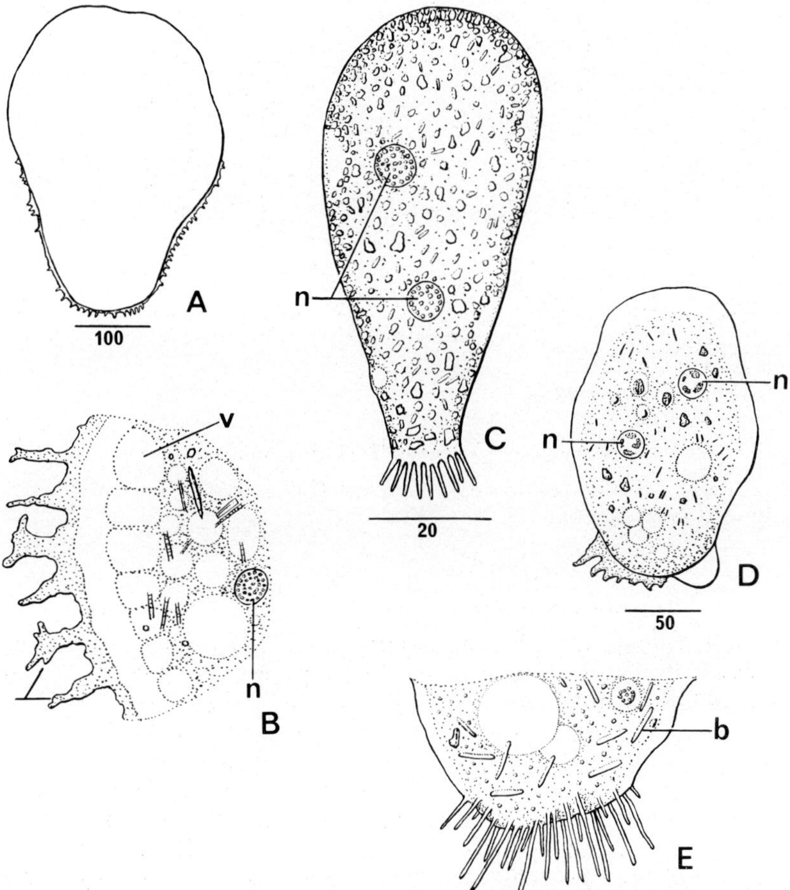

Fig. 9. *Pelomyxa*. A, B, *P. belevskii* (A, outline; B, one edge) (after Penard). C, *P. schiedti* (after Schaeffer). D, E, *P. binucleata* (after Penard). *b*, bacterium; *n*, nucleus; *v*, vacuole.

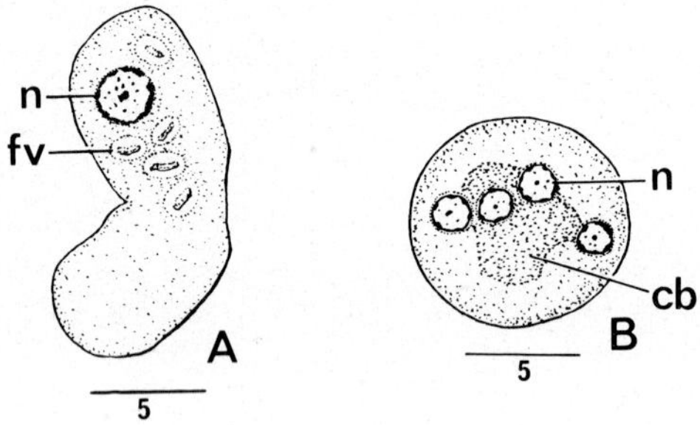

Fig. 10. *Entamoeba moshkovskii*. A, trophic amoeba. B, cyst. (After Neal.) *cb*, chromatoid body; *fv*, food vacuole; *n*, nucleus.

Family ENTAMOEBIDAE Chatton, 1925

Only the one free-living species of this normally parasitic family concerns us here.

Genus ENTAMOEBA Casagrandi & Barbagallo, 1895

All species except the one described here are known to be parasites. Originally uninucleate cyst undergoes nuclear divisions to become 4- or 8-nucleate mature cyst. Nucleus with a single small endosome, central in some species, eccentric in others. Anaerobic.

Isolated from sewage; no host known; growing better at 24°C than at 37°C and not encysting at the higher temperature; amoebae limax-like; mature cysts quadrinucleate, with endosome (sometimes several granules) central, 3–7 rather elongate chromatoid bodies having rounded ends; diameter of amoebae 9–29 μm, generally 11–13 μm; nuclei of fixed amoebae 2·5–4 μm; cysts 6·9–15·3 μm. (Fig. 10)—

Entamoeba moshkovskii Tshalaia, 1941

Soviet Union, England. Morphologically similar to *E. histolytica*. Cysts may be viable up to 300 days at 4°C. Cultivate in HSM+S medium; see page 13.

Family VAHLKAMPFIIDAE Jollos, 1917

Limax amoebae with locomotion by hemispherical, hyaloplasmic eruptions, sometimes alternating to either side at anterior end and sometimes spilling toward posterior along side; nuclear division promitotic; temporary flagellate phase in all genera but one.

Since all members of this family which lack a flagellate stage are classified in the genus *Vahlkampfia*, and since it may be difficult to induce the amoeba-to-flagellate transformation in some species, some now classified in the genus *Vahlkampfia* will probably be found to belong to other genera. This expectation has indeed been fulfilled for one species during the time when this key was being prepared, as noted below.

Induction of transformation is easiest with *Naegleria*. Amoebae from a young and abundant culture (preferably one day old after two or three successive transfers at one-day intervals) are suspended in distilled water, amoeba saline (see p. 11), or certain buffer solutions. Add 15–20 drops of fluid from a pipette to the agar surface (9 cm petri dish), rinse the surface well, and set up hanging drops. These should be kept at 21–28°C for all species except *N. fowleri*, suspected strains of which should be kept at 37°C. A majority of cells of *N. gruberi* and *N. jadini* should transform in 2–4 hours. Observations on suspected *N. fowleri* should continue for 24 hours. *N. thorntoni* is said to transform poorly, with only a few flagellates among thousands of cells in 20–40 hours. Eventually the flagellates of *Naegleria* revert to the amoeboid stage, often on the same day. I always used a known strain of *N. gruberi* as a positive control.

Adelphamoeba galeacystis transforms less well than *Naegleria*, and one must start with a somewhat older culture because of the slow rate of multiplication, otherwise using the same method as for *Naegleria gruberi*. Only a few cells among many in a small quantity of fluid transform.

Obtaining flagellates with *Tetramitus rostratus* is more difficult, and a lowered oxygen concentration, but not completely anaerobic conditions, appears to be important. Attainment of a high population density apparently helps. Fulton (1970) gives the more certain and more complicated method, which also yields more flagellates. I have found that a few flagellates per drop can be obtained using a simpler method. Wash the amoebae off the surface of an abundant agar culture with amoeba saline, as for *Naegleria*, but centrifuge the entire quantity of cell suspension obtained, approximately 7 minutes at or near the highest speed obtained with a small laboratory bench centrifuge, or at about 1250 rpm with a centrifuge of known speed. With a pipette, remove all but 0·2 ml of the supernatant liquid. Add 0·2 ml of proteose peptone-yeast medium, suspend the cells in the liquid, and prepare several hanging drops. Begin observations about 4 hours later on the same day, though usually no flagellates will be seen until late the next day, and then only a few. The temperature range is as for *Naegleria gruberi*. If the first test is negative, try twice more on separate cultures. Proteose peptone-yeast medium consists of 1 g proteose peptone and 0·25 g yeast extract in 100 ml distilled water; it should be autoclaved.

Fulton (1970) gives more information on transformation, including more sophisticated techniques.

Flagellates of the vahlkampfiid genera can be fixed in a wet mount with a drop of saturated mercuric chloride solution for flagellar counts.

Binucleate amoebae and cysts occur more frequently in vahlkampfiids than in most normally uninucleate amoebae, but there seems to be a variation among species.

In connection with the following key to genera, it must be noted that during preparation of this key a flagellate stage was discovered in the species described as *Vahlkampfia jugosa*. The flagellate, which was obtained in most, but not all, strains by the method used for *Naegleria*, has peculiarities which require erection of a new genus for this species. Since the report has not yet been published, the species is retained under the genus *Vahlkampfia* in this key. The user is advised to classify all vahlkampfiids with the amoeboid and cyst characters of *V. jugosa* in that species, whether or not a flagellate stage is found.

1 No flagellate stage— VAHLKAMPFIA

— With flagellate stage— **2**

2 Flagellate stage without cytostome-like (mouth-like) opening; normally biflagellate— NAEGLERIA

— Flagellate stage with cytostome-like opening— **3**

3 Biflagellate, with a small anterior cytostome-like opening near flagella— ADELPHAMOEBA

— Quadriflagellate, flagella originating near beak-like projection (rostrum), posterior to which is cytostome-like opening— TETRAMITUS

Genus VAHLKAMPFIA Chatton & Lalung-Bonnaire, 1912

Limax amoebae moving by means of more or less eruptive, hyaline, hemispherical bulges, with promitotic nuclear division; no flagellate stage.

These amoebae may be considerably broader anteriorly than farther back and often have one or more constrictions along their length Most though not all move rapidly, and the eruptiveness with which most produce their hemispherical bulges during locomotion is so striking in many that little difficulty will be experienced in deciding whether a particular strain is 'more' or 'less' eruptive after a little experience. Some trail a few uroidal filaments produced by adhesion to the substratum. The nucleolus of some species is especially prominent, compared with most vesicular nuclei. Although 'interzonal bodies' (Fig. 1D, E) have never been found in mitotic figures of most species, they appear to occur in *V. debilis*.

The type species *V. vahlkampfi* (Chatton, 1910), originally designated *Amoeba limax* by Vahlkampf (1905), has interzonal bodies in mitosis and has been assumed by recent authors to be a *Naegleria* whose flagellate stage was not seen. However, the cysts figured by Vahlkampf did not resemble those of known *Naegleria* species. *V. vahlkampfi* is not included in this key because an obvious error by Vahlkampf makes it impossible to be certain of its dimensions, but a re-drawn figure of one of Vahlkampf's cysts is included (Fig. 11A).

The genus *Vahlkampfia* as here recognised is equivalent to *Schizopyrenus* Singh, 1952.

Since all well-described freshwater and soil species of the genus produce cysts, cyst structure is used as the first character. Cysts alone are not sufficient for identification of an isolate as a *Vahlkampfia*; the key assumes that generic identification has already been made.

1 Outer wall (ectocyst) of most cysts distinctly separated from inner wall (endocyst) except at some points of contact; not a gelatinous layer— **2**

— Outer wall closely apposed to inner wall in most cysts— **5**

2 Outer wall not wavy, but well separated from inner wall; cyst 8·5–16 μm; 'size' (not defined) of amoebae 20–35 μm. (Fig. 11B, C)— **Vahlkampfia ovis** (Schmidt, 1913)

Sheep feed and faeces; Germany.

— Outer wall often thrown into irregular waves in contact with inner wall at their trough— **3**

3 Cysts often not round; amoebae rather sluggish; 'size' of amoebae 10–20 μm. (Fig. 11D)— **V. hartmanni** (Nägler, 1911)

Fresh water; Austria.

— Cysts usually spherical or ovoid— **4**

4 Maximum cyst diameter *c.* 14 μm, mean 7·5–10·7 μm (depending on strain), most cysts with wavy outer wall; amoebae sometimes with a few uroidal filaments (Fig. 13C)— **V. jugosa** Page, 1967

(Note: A biflagellate phase, not obtained in all strains, has been discovered in this species, which will be made the type species of a new genus.)

Fresh water, soil; Britain, Denmark, U.S.A.

— Cyst diameter greater (to 20 μm); ectocyst often not wavy though widely separated from endocyst (Fig. 11E–G)— **V. russelli** (Singh, 1952)

Soil; Britain.

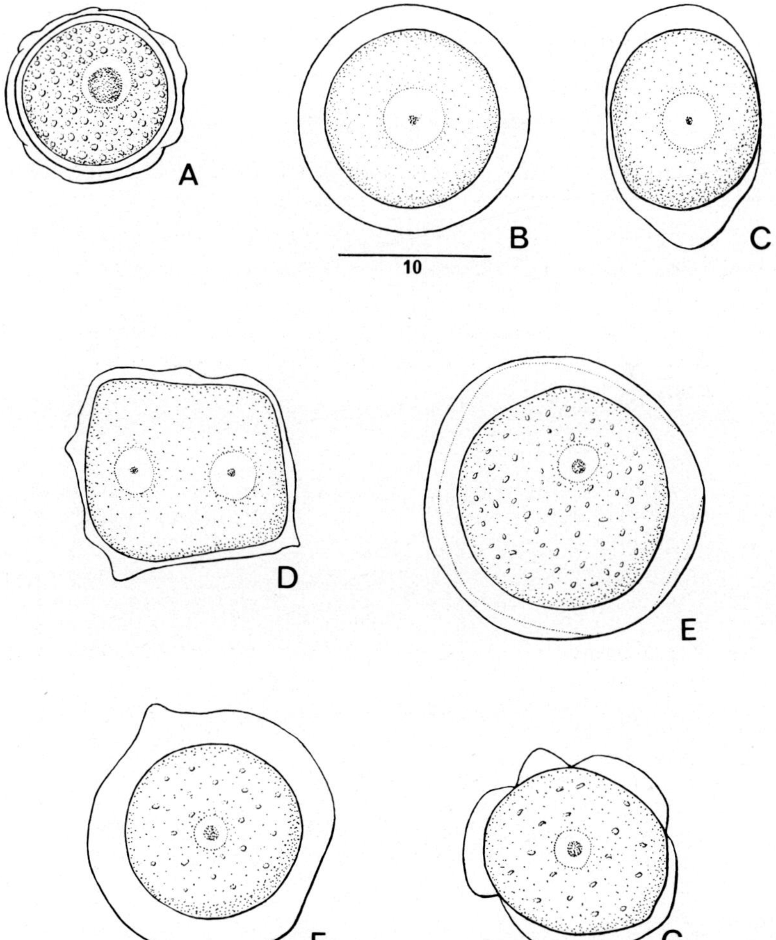

Fig. 11. *Vahlkampfia*, cysts. A, *V. vahlkampfi* (size uncertain) (after Vahlkampf). B, C, *V. ovis* (after Schmidt). D, *V. hartmanni*; binucleate condition atypical (after Nägler). E–G, *V. russelli* (E, after Singh; F, G, after Singh & Das). (C–G to same scale as B.)

5 (1) Cyst with distinct gelatinous coating— **6**

— Cyst without noticeable gelatinous coating (may be somewhat sticky)— **8**

6 Endocyst spherical, with gelatinous layer of irregular thickness; cyst 7–10 μm; average 'size' of amoebae 8–12 μm. (Fig. 12A)— **Vahlkampfia froschi** (Hartmann, 1907)

From tan liquor; Germany.

— Cyst outline circular, oval, slightly angular, or reniform, with outer coating of even thickness though varying from cyst to cyst— **7**

7 Cysts up to 20 μm; cultures may produce red pigment in agar (Fig. 12B)— **V. atopa** (Singh, 1952)

Soil; England.

— Cysts 7–14 μm (mean 9·7 μm); no red pigment observed in agar. (Figs 8D, 13D)— **V. avara** Page, 1967

Fresh water; U.S.A.

8 (5) All or nearly all cysts circular or oval in outline— **9**

— Cysts circular, oval, reniform, pyriform, or irregular in outline— **13**

9 Cysts smooth-walled, averaging about 18–20 μm; amoebae rather sluggish, 20–40 μm rounded up. (Fig. 12C)— **V. magna** Jollos, 1917

Barn manure; Germany.

— Cysts and amoebae otherwise— **10**

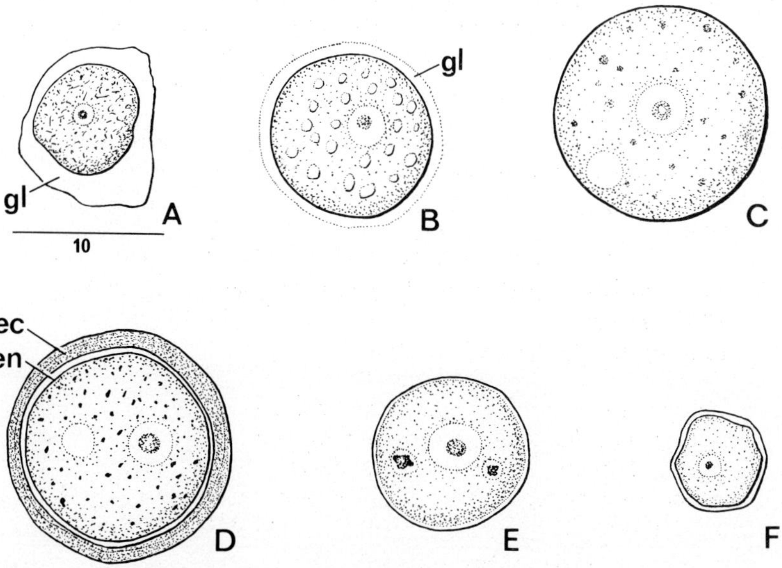

Fig. 12. *Vahlkampfia*, cysts. A, *V. froschi* (after Nägler). B, *V. atopa* (after Singh). C, *V. magna* (after Jollos). D, *V. erythaenusa* (after Singh). E, *V. debilis* (after Jollos). F, *V. lacustris* (after Nägler). (All to same scale as A.) *ec*, ectocyst; *en*, endocyst; *gl*, gelatinous layer.

10 Cysts smooth-walled, with outer layer rather thick; diameter 23 μm; rounded amoebae to 35 μm; pink pigment in agar. (Fig. 12D)— **V. erythaenusa** (Singh, 1952)

Soil; England.

— Cysts and amoebae smaller; no pigment produced— **11**

11 Locomotive form often elongate and straight, not consistently eruptive, 18–37 μm; cysts 6·5–13 μm, with outer coat partly separated in rare cases. (Figs 8E, F, 13E, F)—
V. aberdonica Page, 1974

Soil; Scotland.

— Locomotive forms more consistently eruptive— **12**

12 Amoebae progress rapidly, 'diameter' 15–20 μm; cysts smooth, 10 μm or less; possibly interzonal body in later stages of mitosis. (Fig. 12E)— **V. debilis** Jollos, 1917

Fresh water; Germany.

— Amoebae ball up and change direction frequently, may have a few trailing filaments, 17–42 μm; cysts 8·4–17 μm, outer wall lifted off on one side in minority of cysts; cysts sometimes with vacuolated appearance; multiplies and encysts well at both 23°C and 37°C (Figs 8G, 13G)— **V. enterica** Page, 1974

Intestine of turkey; Czechoslovakia.

Fig. 13. *Vahlkampfia*. A, locomotive form of *V. ustiana*. B, locomotive form of *V. inornata*. C–G, cysts. C, *V. jugosa*. D, *V. avara*. E, F, *V. aberdonica*. G, *V. enterica*. (A, B, ×1,000; C–G, ×2,000.)

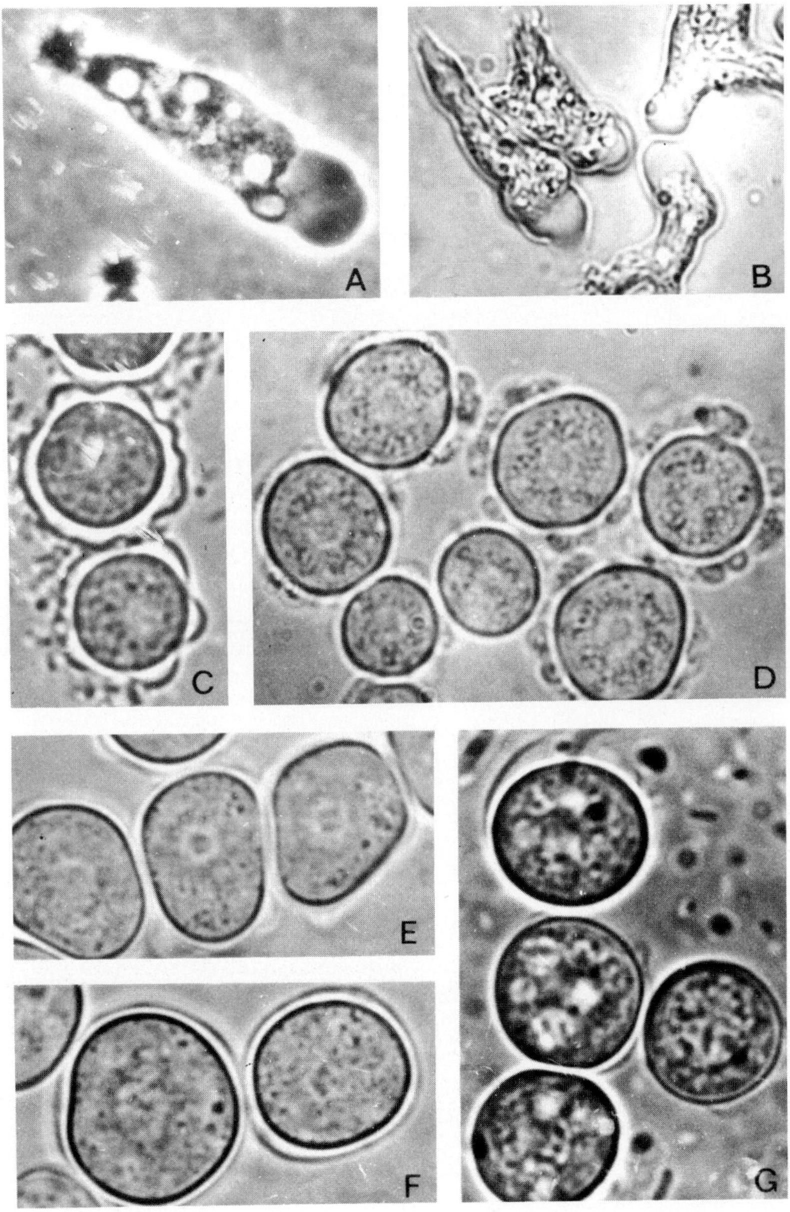
A
B
C
D
E
F
G

13 (8) Amoebae often with prominent adhesion uroid on bulbous posterior end (Fig. 13A), 30–65 μm; cyst outline circular, oval, reniform, pyriform; cysts usually 12·5–21 μm excluding outer coat, which may be sticky; multinucleate irregularly shaped cysts not uncommon; polar masses with somewhat fragmented appearance at some stages of mitosis. (Figs 13A, 14A, B)— **V. ustiana** Page, 1974

Fresh water: Czechoslovakia.

— No prominent uroidal tuft— **14**

14 Cyst outline circular, oval, reniform, with thin, somewhat sticky outer layer, 8·6–13·2 μm; amoebae commonly with a few uroidal filaments, 18·6–36·6 μm; polar masses compact, not fragmented. (Figs 13B, 14C)— **V. inornata** Page, 1967

Fresh water; U.S.A.

— Cyst shape irregular, poorly described, 6–8 μm; amoebae seldom extended, changing form rapidly, 'size' (undefined) 8–15 μm. (Fig. 12F)— **V. lacustris** (Nägler, 1909)

Fresh water; Austria.

Fig. 14. A, B, *Vahlkampfia ustiana*, cysts. C, *V. inornata*, cysts. D, E, *Naegleria gruberi*, amoebae. F, *N. gruberi*, flagellate fixed with $HgCl_2$. (A, D, F, ×1,000; B, ×1,500; C, ×2,000; E, ×400.)

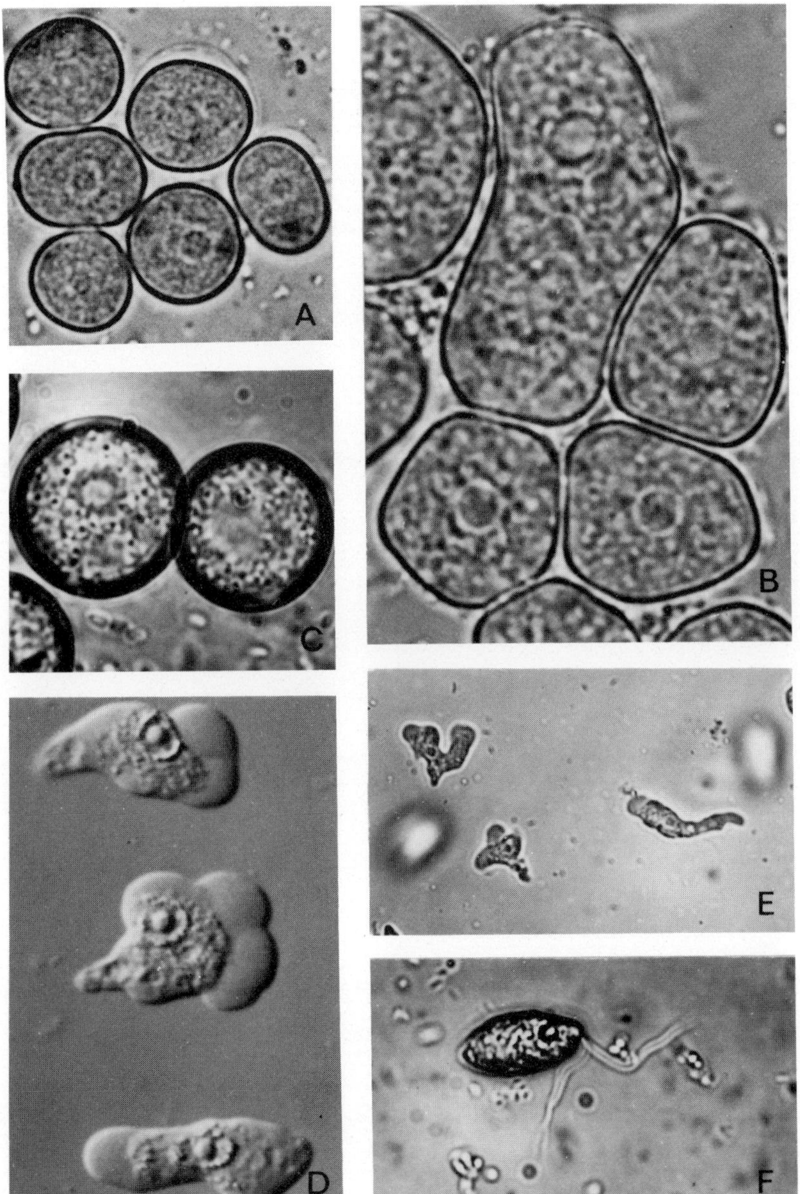
A
B
C
D
E
F

Genus Naegleria Alexeieff, 1912; emend. Calkins, 1913

Limax amoebae moving by means of more or less eruptive, hyaline, hemispherical bulges; promitotic nuclear division; temporary flagellate stage, usually with two flagella of equal length, lacking cytostome (mouth-like opening), with nucleus in anterior region.

The genus as here defined includes *Didascalus* Singh, 1952.

The plugged pores in the cyst walls of *N. gruberi* can be seen in living cysts with ordinary bright-field microscopy, though those of the 'smooth' (by far the most common) strains are more distinct with differential interference contrast (Fig. 16B, C). For careful identification the simple periodic acid-Schiff's procedure (Galigher & Kozloff, 1971; Grimstone & Skaer, 1972) is recommended. The presence of pores in *N. fowleri* is barely suggested by PAS staining (which actually demonstrates the plugs rather than the pores) and was conclusively demonstrated only by scanning electron microscopy. A recently published photograph (Jonckheere & Van de Voorde, 1975) shows empty pores of an empty *N. fowleri* cyst in side view, using phase contrast, but most workers have failed to see the pores in this species. The major reason for the superior visibility of the *N. gruberi* pores is the presence of an external rim or collarette, lacking in other species.

For information on the involvement of *Naegleria* in the lethal disease, primary amoebic meningo-encephalitis, the reader is referred to the many recent papers on that topic (e.g., Carter, 1972; Jadin, 1973; Symmers, 1969). At the time of writing, the majority of workers consider only *N. fowleri* a lethal pathogen, but *N. gruberi* is not yet entirely in the clear.

See suggestions on handling suspected pathogens on page 20.

This key employs only morphological and certain physiological distinctions, but antigenic differences have been discovered and may eventually be the main method for clinical differentiation.

1 Cysts spherical to ovoid, usually smooth-walled, but with thin outer wall separated in some (rough) strains which often have a more irregular over-all form; several pores containing plugs discernible with light microscope (prominent in rough strains); usually a perinuclear layer of coarse granules in cysts, often in amoebae; cysts 9–23 μm, mean 10–16 μm (according to strain); most strains transforming readily to flagellates between 20°C and 25°C. (Figs 14D–F, 16A–F)—

Naegleria gruberi (Schardinger, 1899)

Possibly the most common freshwater amoeba and apparently world-wide in occurrence.

— Distinct plugs not discernible in unstained cysts with light microscope— **2**

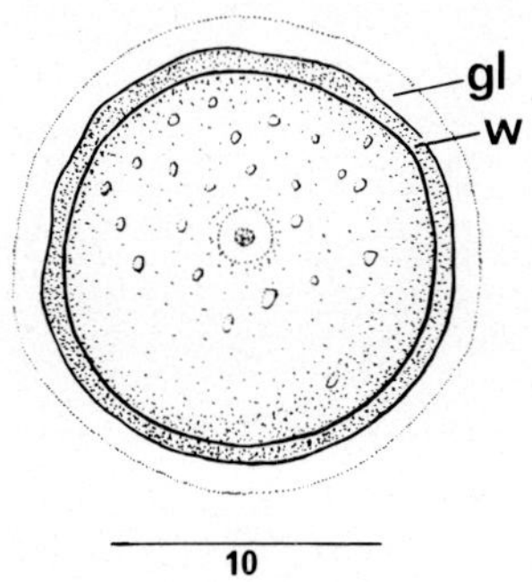

Fig. 15. *Naegleria thorntoni*, cyst (after Singh). *gl*, gelatinous layer; *w*, cyst wall.

2 Cysts smooth-walled (approximately 11–21 μm) with fairly thick gelatinous outer layer; no perinuclear layer of granules reported; no interzonal body in mitosis; few flagellates produced in 20–40 hours at 20–21°C. (Fig. 15)—

N. thorntoni (Singh, 1952)

Soil; England.

— No thick gelatinous coating on cysts; interzonal bodies usually found in mitotic figures; perinuclear layer of granules sometimes present— **3**

3 Cysts spherical unless clumped closely together, 7–15 μm; nucleus may be obscure in some cysts but usually easily distinguishable; growth and transformation of flagellates much better at 37°C than at 20–25°C; acutely pathogenic to mice by nasal instillation. (Fig. 16G)— **N. fowleri** Carter, 1970

Identified as the principal causative agent of primary amoebic meningo-encephalitis; reported sporadically from most continents and almost certainly present in Britain; favoured by water temperatures in 30's and possibly even mid-20's, that is, temperature of heated swimming pools. This diagnosis accepts the synonymy of *N. aerobia* Singh & Das, 1970, with *N. fowleri*, which is disputed by Singh & Das, 1972, but supported by recent immunological work (Visvesvara & Healy, 1975).

— Cysts spherical, 9–18 μm; grows well at 28°C but poorly at 37°C; not acutely pathogenic to mice by nasal instillation. (Fig. 16H)— **N. jadini** Willaert & Le Ray, 1973

Swimming pool; Belgium.

Fig. 16. *Naegleria*, cysts. Plugged pores indicated by arrows. A–C, *N. gruberi*, smooth cysts; C, surface view with differential interference contrast. D, E, *N. gruberi*, rough cysts. F, *N. gruberi*, angular but fairly smooth cysts with thick plugs. G, *N. fowleri* (strain HB–1, also designated as *N. aerobia*). H, *N. jadini* (F, ×1,200; all others, ×2,000.)

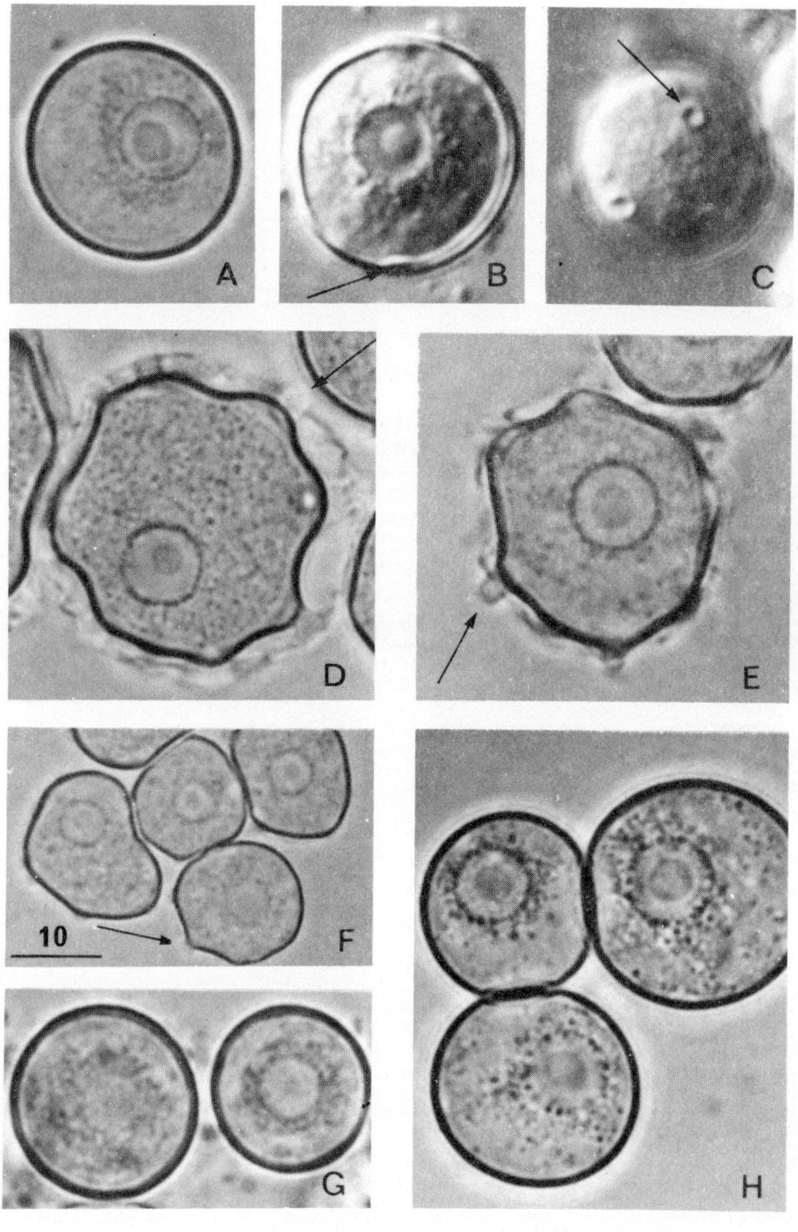
A
B
C
D
E
10
F
G
H

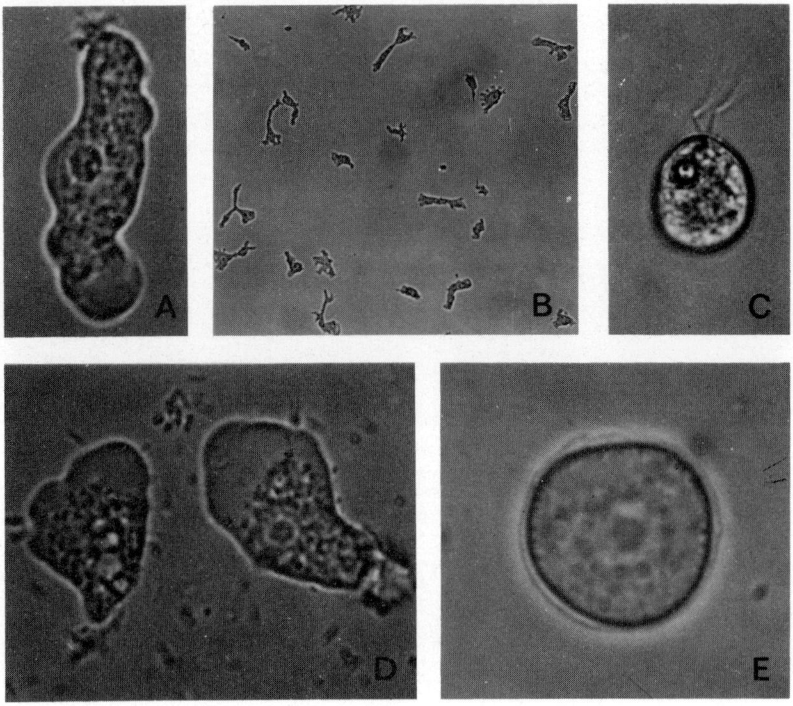

Fig. 17. A-C, *Adelphamoeba galeacystis*. A, B, amoeboid phase. C, flagellate phase. D, E, *Tetramitus rostratus*. D, amoebae. E, cyst. (A, C, D, ×1,000; B, ×100; E, ×2,000.)

Genus ADELPHAMOEBA Napolitano, Wall & Ganz, 1970

Limax amoebae moving by more or less eruptive, hyaline, hemispherical bulges, with promitotic nuclear division, and with a temporary, biflagellate swimming stage with simple oval cytostome. One described species.

Amoebae 11·8–30·5 μm, may be extended and branched, but with more compact, vahlkampfiid limax form in active locomotion; often trailing several to many uroidal filaments; flagellate 9·0–18·8 μm, oval to pyriform, with nucleus always in anterior region; cysts usually rounded with distinctive brim reflected outward at one pole, giving helmet-shaped appearance, with deep depression inside brim, 8–11 μm. (Figs 17A–C, 18)—

Adelphamoeba galeacystis
Napolitano, Wall, & Ganz, 1970

Soil; U.S.A.

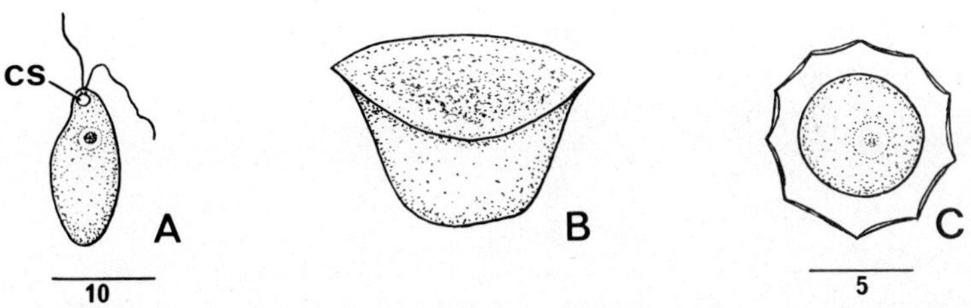

Fig. 18. *Adelphamoeba galeacystis.* A, flagellate. B, helmet-shaped cyst. C, cyst in transverse view. (After Napolitano, Wall & Ganz; B to same scale as C.) *cs*, cytostome.

Genus TETRAMITUS Perty, 1852

Limax amoebae moving by more or less eruptive, hyaline, hemispherical bulges, with promitotic nuclear division and with a normally quadriflagellate swimming stage with a cytostome.

Several species have been assigned to this genus, but only one non-marine species is known to have an amoeboid phase. The possibility of confusion with *Vahlkampfia* when flagellates are not seen was pointed out (p. 51). Since the connection between the amoeboid and flagellate stages was recognized in 1922, *T. rostratus* has been isolated only from faeces, the posterior digestive tracts of rats and insects, and (in one case) human urine. There are, however, good indications that *Tetramitus* is to be found in soil and water, flagellates having been reported from aquatic habitats. For further information on the genus and on transformation, see Fulton (1970).

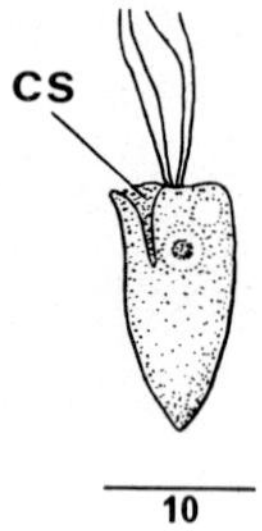

Fig. 19. *Tetramitus rostratus*, flagellate. *cs*, cytostome.

Amoebae 16–30 μm, sometimes spreading anteriorly to a markedly triangular form; flagellate 14–18 μm, with four equal flagella and a cytostomal groove which is broad anteriorly, with a distinct rostrum; promitotic nuclear division accompanying reproduction in both amoeboid and flagellate stages; cysts 6–18 μm, spherical or ovoid, with smooth inner wall and thinner outer wall which is often raised slightly in one or two places. (Figs 17D, E, 19)— **Tetramitus rostratus** (Perty, 1852)

Family HARTMANNELLIDAE Volkonsky, 1931; emend. Page, 1974

This family includes all truly limax amoebae (and only limax amoebae) with a vesicular nucleus and mesomitotic nuclear division.

It must be emphasized that presence or absence of the hyaline cap is not absolute but varies with time in a given cell; no difficulty should be experienced in using this character with a fair number of observations.

1 Hyaline cap almost always present in continuing locomotion, usually at least as deep antero-posteriorly as broad; no nuclear division in cysts, if cysts formed— HARTMANNELLA

— Hyaline cap often or usually absent— **2**

2 Cysts often binucleate (also uninucleate, rarely trinucleate) as result of mitosis within cyst, often containing one, rarely two, densely staining bodies; hyaline cap fairly common in locomotion though often obliterated by intrusion of granules— GLAESERIA

— No nuclear division in cysts, if cysts formed; hyaline cap usually present only at initiation of pseudopodium or as shallow crescent— 3

3 Villous-bulb uroid often present; nuclear membrane usually very distinct— SACCAMOEBA

— Villous-knob uroid never formed— CASHIA

Genus HARTMANNELLA Alexeieff, 1912; emend. Page, 1974

Limax amoebae moving by steady flow of cytoplasm, with hyaline cap practically always present in continuing locomotion, this hyaline cap usually at least as deep antero-posteriorly as broad; no villous-bulb uroid or extensive filamentous adhesion uroid; in some, a few occasional trailing filaments or small morulate uroid. Mitosis with early disintegration of nucleolus and nuclear membrane. No nuclear division in cysts, if cysts are formed.

Perhaps 30 or more species have been attributed to this genus, but most such generic designations are meaningless because of the past vagueness of the generic diagnosis. Nevertheless, more described species than the two below could probably be included if they were re-investigated. Some recently described species have been transferred to other genera, others cannot be included in the key at all because of inadequate descriptions, and two parasitic species have been omitted. Some species which have in the recent past been included under an imprecisely defined genus *Hartmannella* may be found under *Acanthamoeba*, *Cashia*, *Echinamoeba*, and *Saccamoeba*.

1 L/B usually about 4·5, maximum about 6·5; hyaline zone, as deep as broad, usually present, occasionally obliterated; one contractile vacuole, usually several small crystals; cysts smooth, spherical or ovoid, with two layers closely apposed, refractile cytoplasmic granules; amoeba 16–33 μm, cyst 7–13 μm. (Fig. 20A, C)— **Hartmannella cantabrigiensis** Page, 1974

Fresh water; England.

— L/B usually more than 6, maximum about 8·4; hyaline zone always present, depth to twice breadth; usually more than one contractile vacuole; cysts spherical or slightly ovoid, with ectocyst somewhat separated in minority; amoeba 12–33 μm, cyst 4–9·5 μm. (Fig. 20B, D)— **H. vermiformis** Page, 1967

Fresh water, leaf litter; England, U.S.A., Sweden.

Genus Glaeseria Volkonsky, 1931

Limax amoebae moving generally by steady flow, with no explosively produced bulges; hyaline cap present at initiation of locomotion and sometimes during continued locomotion; no villous uroid; mesomitosis; nuclear division also during early cystic stages, with mature cysts often binucleate and sometimes containing a dense body slightly larger than nucleus.

Only one species known.

Slender, with L/B averaging more than 4 and sometimes exceeding 6·5; sometimes with uroidal knob and occasionally a few uroidal filaments; cyst with smooth inner wall, more or less smooth outer wall, and sticky coating; amoeba 20–46 μm, cyst 7·4–19 μm. (Fig. 20E–J)— **Glaeseria mira** (Gläser, 1912)

Fresh water; Germany, England.

Fig. 20. A, B. *Hartmannella*, amoebae. A, *H. cantabrigiensis*. B, *H. vermiformis*. C, D, cysts of (C) *H. cantabrigiensis* and (D) *H. vermiformis*. E–J, *Glaeseria mira*. E–G, locomotive forms. H, cyst with especially prominent gelatinous coat. I, cysts showing nucleus (*n*) and darkly staining or dense body (*db*). J, cyst in haematoxylin preparation. (A, B, E–G, ×1,000; C, D, H–J, ×2,000.)

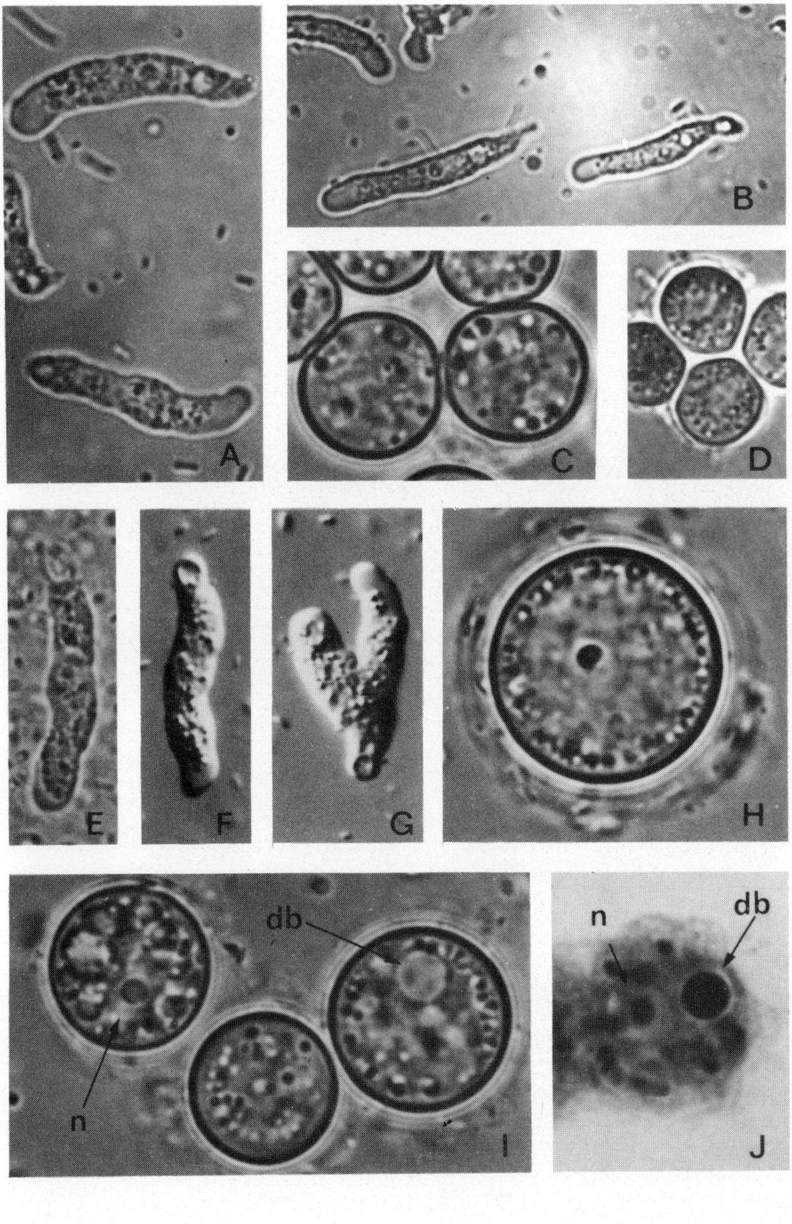
A
B
C
D
E
F
G
H
db
n
I
n
db
J

Genus Saccamoeba Frenzel, 1892; emend. Bovee, 1972

Limax amoebae with hyaline cap reduced to shallow crescent or absent in continuing locomotion, though always seen on initiation of pseudopodium; advance by steady flow, often with non-eruptive slight antero-lateral bulging; villous-bulb uroid with more or less rigid fine villi often present; nuclear membrane usually distinct in living specimens; mitosis with early disintegration of nucleolus, nuclear membrane often persisting through metaphase.

Since many amoebae in a preparation may be inactive, it is possible in a cursory observation to miss the characteristic limax form. Furthermore, the amoebae may be transiently polypodial when initiating locomotion or changing direction. The characteristic villous-bulb is not present in all cells at all times, and the villi are quite short in some species. The bulging of the contractile vacuole at the posterior end or just anterior to the uroidal bulb is characteristic, though not exclusive to this genus.

An amoeba described in the key as 'broad' does not attain a length: breadth ratio of 4, while a 'slender' one does. This character refers to monopodial amoebae in fairly rapid locomotion; the lengths are also for monopodial amoebae.

The species isolated from English soils and described under the names *Amoeba agricola* Goodey, 1916; *Hartmannella agricola*, identified by Singh with Goodey's species; and *Hartmannella leptocnemus* Singh, 1952, may belong to this genus (less likely, to *Cashia*) but the published descriptions do not include details necessary for a decision.

1 Cyst-forming; no distinctly crystalline cytoplasmic inclusions in amoebae; cyst with circular or slightly oval outline, usually distinct sticky outer coat; amoeba 30–75 μm, with nucleus 4·6–8·4 μm; cyst 12–19 μm. (Fig. 21A–C)—

Saccamoeba stagnicola Page, 1974

Fresh water; England.

— No cysts known; with distinct cytoplasmic crystals— **2**

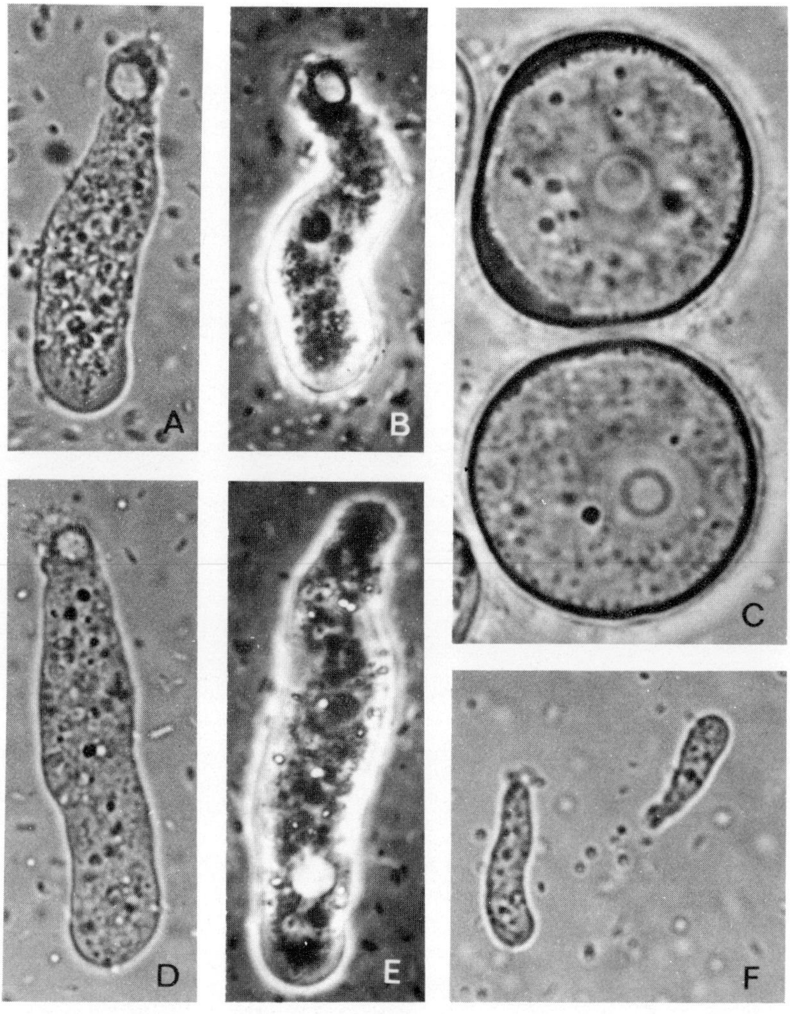

Fig. 21. A–C, *Saccamoeba stagnicola*, locomotive forms and cysts. D, E, *S. limax*, locomotive forms. F, *Cashia limacoides*, locomotive forms. (A, B, D–F, ×1,000; C, ×2,000.)

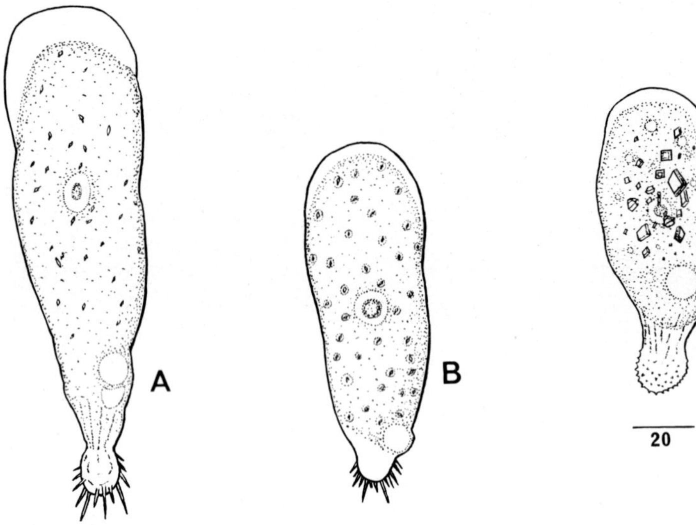

Fig. 22. *Saccamoeba.* A, *S. wakulla.* B, *S. limna.* C, *S. lucens.* (All after Bovee; all to same scale as C.)

2 Numerous small (1–3·5 μm) crystals; amoebae broad— **3**

— Usually fewer than 25 crystals; amoebae broad or slender— **4**

3 Up to 175 μm but some not exceeding 130 μm; pseudopodia in slow locomotion sometimes flattened; distinct hyaline cap in rapid advance; wrinkled villous-bulb uroid; nucleus 3·5–6·5 μm; one to several contractile vacuoles. (Fig. 22A)—
S. wakulla Bovee, 1972

Fresh water; U.S.A.

— Up to 120 μm; hyaline cap reduced in rapid advance; uroidal villi on clear disc or bulb; nucleus 10–15 μm; one contractile vacuole (Fig. 22B)— **S. limna** Bovee, 1972

Fresh water; U.S.A.

4 (2) Broad, 70–100 μm; uroid smooth or with punctate appearance; nucleus 6–7 μm; usually fewer than 20 crystals, 5–15 μm, often appearing cuboidal or plate-like. (Fig. 22C)—
S. lucens Frenzel, 1892, sensu Bovee, 1972

Fresh water; U.S.A., possibly Argentina and Switzerland.

— Slender (mean L/B 5·5), 40–75 μm; villous bulb common; nucleus 5·8–8·6 μm; truncate bipyramidal crystals to 2·5 μm or longer, seldom more than 30 present, occasionally none. (Fig. 21D, E)— **S. limax** (Page, 1969)

Fresh water, soil; U.S.A., Britain, possibly elsewhere in Europe.

Genus CASHIA Page, 1974

Limax amoebae with hyaline cap usually absent or reduced to slight crescent except upon initiation of new pseudopodium; advance by steady flow, occasionally with non-eruptive slight antero-lateral bulging; no villous-bulb uroid or extensive filamentous adhesion uroid; nuclear division mesomitotic in one species investigated; no cysts known.

1 Length 7·5–25·5 μm, tapering from rather broad anterior end (mean L/B 3·3) to narrow posterior end, which may be smooth or bear small bulb; crescentic hyaline cap much reduced when present; lateral pseudopodium occasionally but not often produced to change direction; nucleus 2·8–4·1 μm; several coarse cytoplasmic granules (Fig. 21F)—
Cashia limacoides (Page, 1967)

Fresh water; U.S.A., Czechoslovakia.

Fig. 23. *Cashia angelica* (after Bovee).

— Length 50–100 μm, slender in rapid locomotion; crescentic hyaline cap usually present; morulate uroid common; lateral pseudopodia appearing in slow locomotion; nucleus 5–6 μm; crystals. (Fig. 23)— **Cashia angelica** (Bovee, 1972)
Fresh water; U.S.A.

Family COCHLIOPODIIDAE De Saedeleer, 1934

This family, which some workers consider intermediate between naked and testate amoebae, is poorly defined, and the generic position of some included species is rather uncertain. It is included in this key because some species of *Cochliopodium* and at least one of *Gocevia* would be taken for naked amoebae by anyone seeing them for the first time.

The original diagnosis of *Gocevia* separated it from *Cochliopodium* by occurrence of foreign bodies in the hyaline cuticle of the former, but these are absent from the newest and best-described species. A centrosphere has been reported in one species of each genus.

1 Central or post-central granular mass surrounded by rather narrow, flattened hyaloplasmic veil; in locomotion, granular mass of some species merely the hump-like centre of a disc; in other less common species, it is more globose; flexible, closely adherent tectum (covering) on dorsal surface, often appearing as double-lined layer (Fig. 26A) when granular hump is seen in profile— COCHLIOPODIUM

— Granular mass globular or hump-like, preceded but not entirely surrounded by flattened hyaline border producing short projections; hyaline, cuticle-like covering somewhat less flexible than tectum of *Cochliopodium* and at times separated from protoplast— GOCEVIA

Genus COCHLIOPODIUM Hertwig & Lesser, 1874; emend. Bark, 1973

In locomotion a raised central or post-central granular mass, surrounded by a thin hyaloplasmic veil; when inactive, spherical or bell-shaped; floating form often radiate with slender, conical hyaline pseudopodia. Flexible tectum or covering of small scales closely applied to much of cell surface, restricted to dorsal surface in locomotion and not extending over outermost zone of hyaline border. This tectum may appear simply as a doubly-contoured coat when seen on the hump in side view (Fig. 26A) or as faint or distinct punctation over the hyaline zone (Fig. 25–AC); details of the scales are distinguishable only by electron microscopy. The cytoplasm of the common species contains prominent crystals.

Some of the less flattened species have also been classified in other genera of testate amoebae. In general the species diagnoses show a need of re-investigation.

Distinguish carefully from *Hyalodiscus*.

1 Spines in tectum visible with light microscope; usually or often globose— **2**

— Without spines visible with light microscope; usually disc-like in locomotion— **3**

2 Spines 23–29 μm long; diameter of rounded cell *c.* 42 μm; one contractile vacuole or sometimes more. (Fig. 24A)—
Cochliopodium longispinum West, 1901

Fresh water; England.

— Spines shorter; average diameter of cell 35 μm, may be only 20 μm; may take flattened form with scales visible over hyaline border; apparent zoochlorellae; 3 or more contractile vacuoles (Fig. 24B)— **C. vestitum** (Archer, 1871)

Fresh water; Europe, U.S.A.

3 (1) With adherent debris or amorphous grains on tectum— **4**

— Without adherent material on tectum— **5**

4 Granular mass a high dome even in expanded locomotive form; radial lines of punctation on tectum; occasionally a free space between cell and tectum; 70–120 μm or more. (Fig. 24C)—
C. granulatum Penard, 1890

Fresh water; Europe.

— More flattened and disc-like in locomotion; no punctation visible on very flexible tectum; heavy covering of refractile grains on tectum; 25–30 μm rounded, up to 50 μm flattened, not including fine pseudopodia. (Fig. 24D)— **C. obscurum** Penard, 1890

Fresh water; Europe.

5 (3) No punctation visible on tectum; hyaline veil often producing finely tapering pseudopodia; diameter of tectum 12·4–13·5 μm. (Fig. 24E, F)— **C. minutum** West, 1901

Fresh water; Wales.

— Larger; flattened locomotive form with punctation of tectum (scales) visible easily or with difficulty; L/B usually less than 1; few or many conspicuous truncately bipyramidal crystals; nucleus usually surrounded by layer of refractile granules— **6**

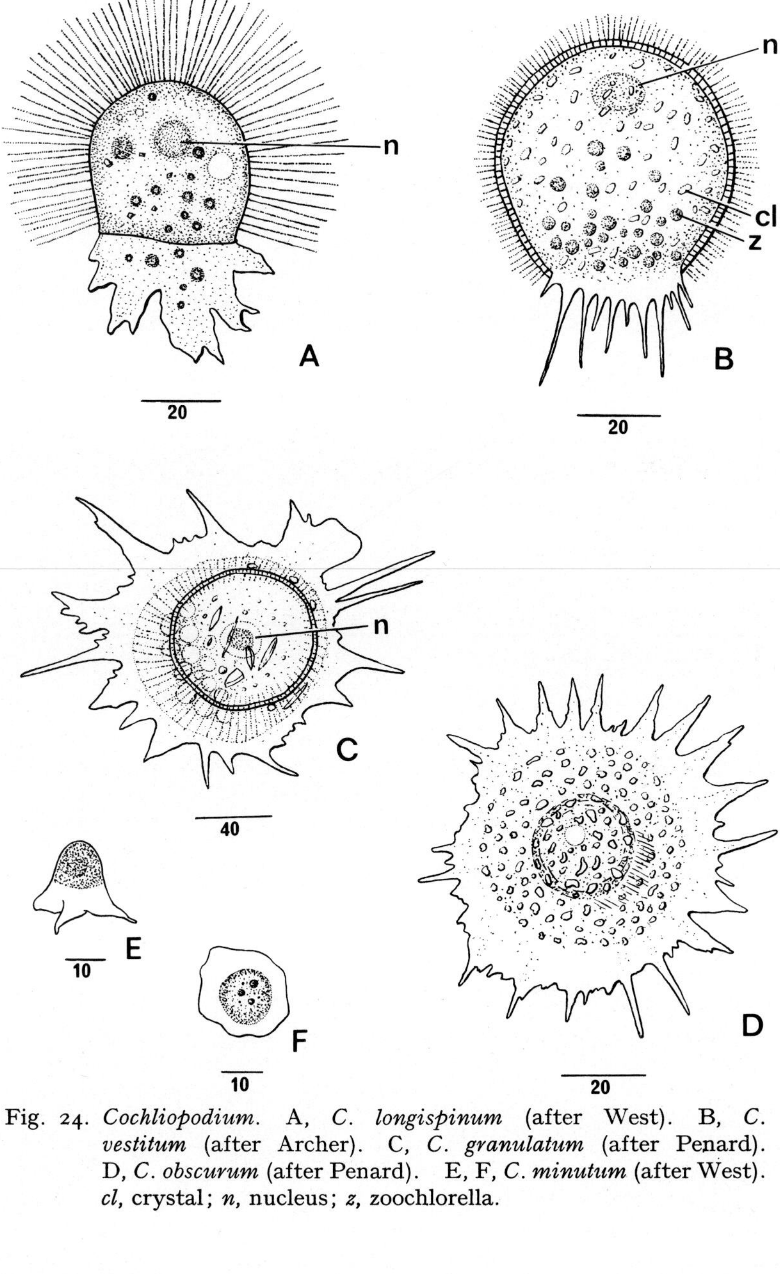

Fig. 24. *Cochliopodium*. A, *C. longispinum* (after West). B, *C. vestitum* (after Archer). C, *C. granulatum* (after Penard). D, *C. obscurum* (after Penard). E, F, *C. minutum* (after West). *cl*, crystal; *n*, nucleus; *z*, zoochlorella.

6 Scales clearly visible with light microscope; distinct tubular configuration of tectum around pseudopodial bases of floating form; finely conical, sometimes furcate pseudopodia often produced from anterior margin; usually 100–300 cytoplasmic crystals; greater diameter of locomotive form 49–90 μm; nucleus 11–12·5 μm. (Fig. 25A, B)—
C. bilimbosum (Auerbach, 1856)

Fresh water, sewage sludge; Europe, U.S.A.

— Punctation visible with difficulty; amoebae generally smaller than above; anterior border seldom or never producing distinct pseudopodia— 7

7 Locomotive form 14–80 μm, usually 24–55 μm; tectum clearly visible as doubly-contoured covering on granular hump; pseudopodia rarely if ever produced by anterior edge; often many trailing filaments; nucleus 6·9–13·8 μm; crystals sometimes numbering 30 or more; cyst spherical or oval, with smooth inner wall and thick, irregular outer layer, 18·5–27 μm. (Figs 25C; 26A, B)— **C. actinophorum** (Auerbach, 1856)

Fresh water; Europe, U.S.A.

— Locomotive form 14·5–38·5 μm, usually 19–38 μm; short, finely conical pseudopodia sometimes produced by anterior edge; sometimes with trailing filaments along posterior edge; nucleus 4·8–6·9 μm; crystals numbering up to 10; cyst lacking in some strains, present in others, 11·5–21 μm. (Fig. 26C, D)—
C. minus sp. nov.

Cochliopodium minus sp. nov.

Diagnosis: Disc-like in locomotion, with breadth usually greater than length; anterior hyaloplasmic border usually a broad arc with slight irregularities, sometimes producing short, finely conical pseudopodia; posterior hyaloplasmic border sometimes trailing short filaments; nucleus usually surrounded by a single layer of yellowish, highly refractile granules; nucleolus usually indistinct in living

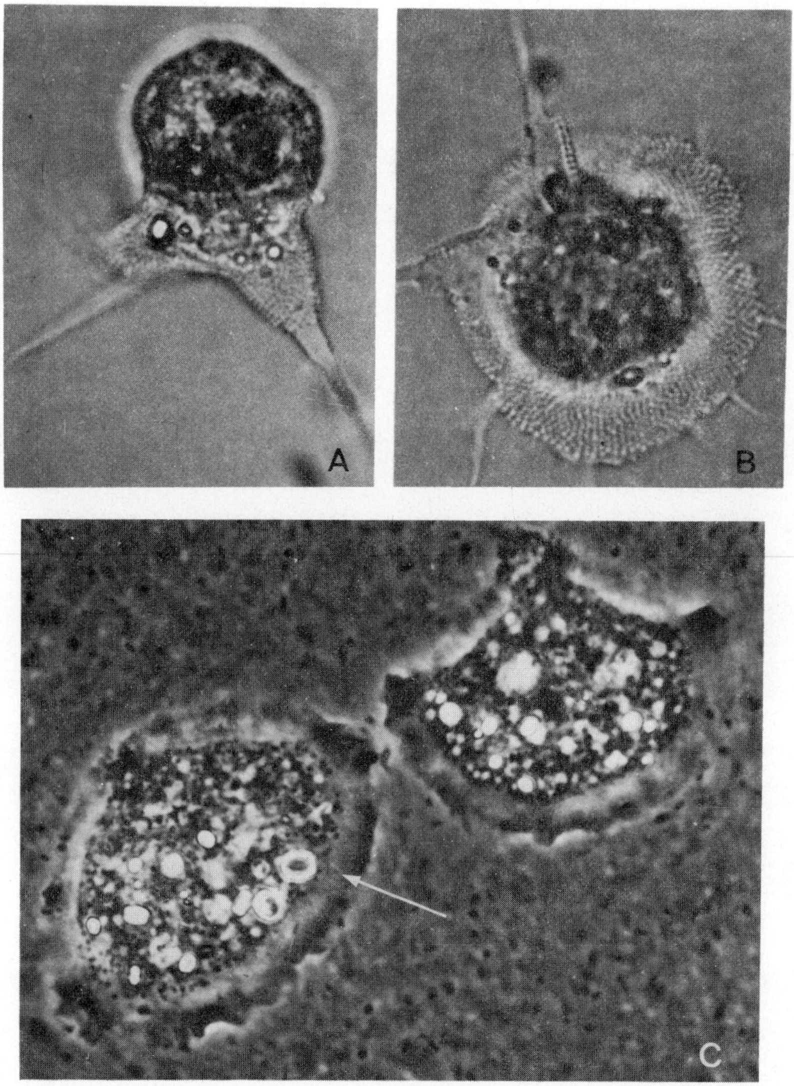

Fig. 25. *Cochliopodium*. A, B, *C. bilimbosum*. C, *C. actinophorum*, with scales visible as punctation (arrow) on hyaline border of amoeba at left. (All × 1,000.)

specimens, occasionally in two pieces; several contractile vacuoles; usually a number of truncately bipyramidal cytoplasmic crystals, numbering up to 10, approximately 1 μm long; greatest dimension of locomotive form 14·5–38·5 μm, usually 19–38 μm; diameter of nucleus 4·8–6·9 μm; cysts produced by some strains, 11·5–21 μm with smooth inner wall and thick, somewhat uneven, outer wall.

Observed habitats: Fresh water, England, U.S.A.; fairly common.

(This species is equivalent to '*Hyalodiscus actinophorus* var. *minor*' described by Page in 1968; varieties have no standing in zoological nomenclature.)

Genus Gocevia Valkanov, 1932

Hyaline cuticle-like covering which may at times be separated from the protoplast and is somewhat less flexible than the tectum of *Cochliopodium*. Granular mass globular or flattened and hump-like, preceded by a flattened hyaline border with a ragged anterior edge bearing short conical or pyramidal projections.

1 Hyaline cuticle without foreign bodies, distinguished with some difficulty; protoplast filling cuticle completely; uninucleate; unornamented cyst; amoeba generally below 50 μm; cyst 12–16 μm. (Fig. 26E)— **Gocevia fonbrunei** Pussard, 1965

— Usually with covering of foreign bodies, though young cells may be nearly free of foreign bodies; protoplast not filling cuticle completely; usually or always binucleate; several contractile vacuoles; 29–30 μm. (Fig. 27)—
G. binucleata De Saedeleer, 1934

Fig. 26. A, *Cochliopodium actinophorum*, with tectum distinguishable as double line indicated by arrow. B, cyst of *C. actinophorum*. C, D, *Cochliopodium minus*. E, *Gocevia fonbrunei* (photograph by Dr M. Pussard). (A, C, D, ×1,000; B, ×1,200; E, ×1,440.) *ce*, centrosphere; *n*, nucleus.

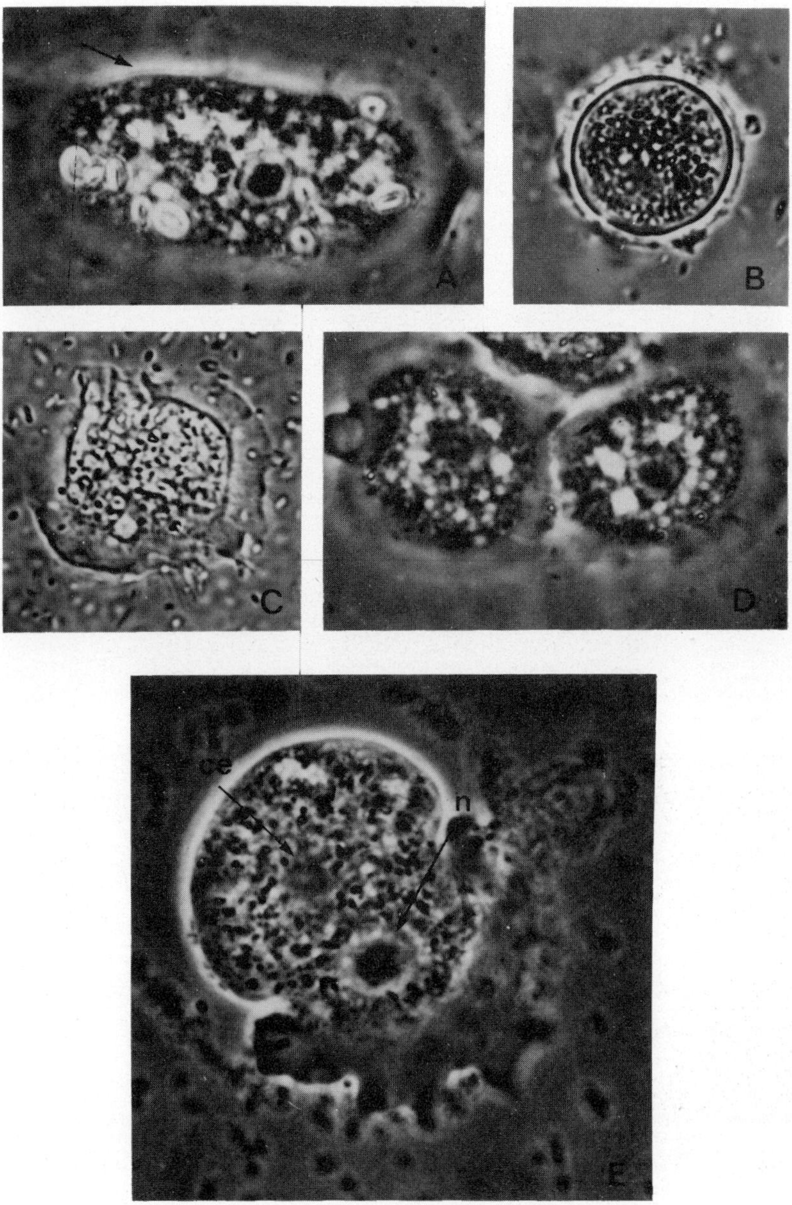
A
B
C
D
ce
n
E

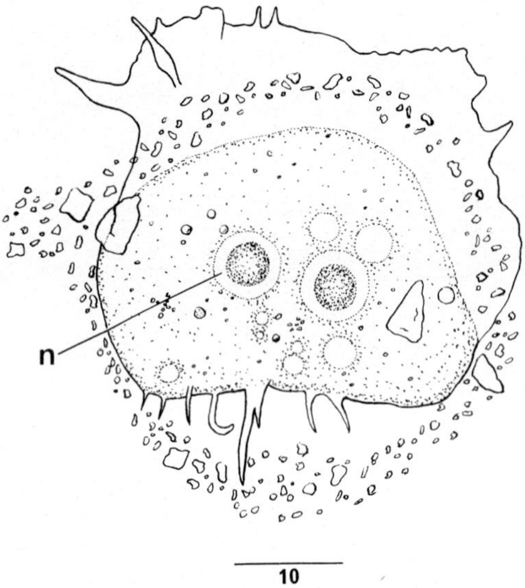

Fig. 27. *Gocevia binucleata* (after De Saedeleer). *n*, nucleus.

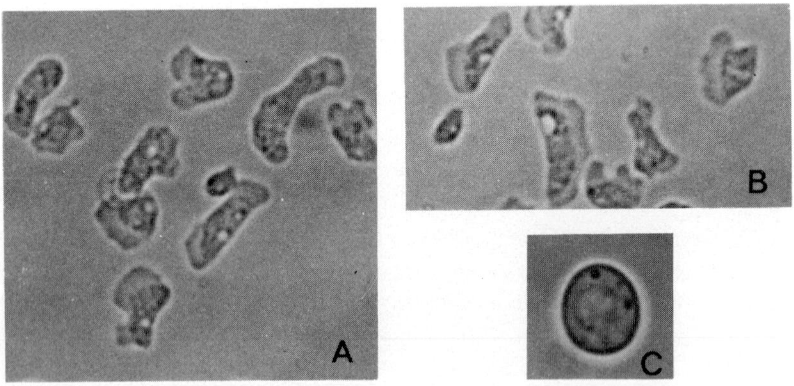

Fig. 28. *Rosculus ithacus*. A, B, active amoebae with variety of forms (×1,000). C, cyst (×2,000).

Family FLABELLULIDAE Bovee, 1970

No certain members of the genus *Flabellula* as re-defined by Bovee have yet been described from habitats other than salt water. The genus *Vannella*, which Bovee separated from *Flabellula*, is here classified in the Thecamoebidae, and *Flamella* in the Hyalodiscidae.

Genus Rosculus Hawes, 1963

Small amoebae with rapidly changing form, varying from greatly expanded laterally to temporarily elongate, sometimes spatulate or flabellate, often asymmetrical; hyaline zone usually with smoothly irregular edge, sometimes deeply divided; no spiny pseudopodial projections; cytoplasmic flow often eruptive, with hyaline wave sometimes running around periphery of granular mass; nuclear division with disintegration of nucleolus and nuclear membrane.

One known species.

Greatest dimension 5–17 μm, exceptionally 25 μm; nucleus 1·9–2·4 μm; contractile vacuole reported for two strains, not for third; cysts smooth, oval or spherical, 3·5–8·8 μm. (Fig. 28)—
Rosculus ithacus Hawes, 1963

Fresh water, leaf litter, rectum of *Natrix natrix*, on frozen maize; Britain, U.S.A.

Family THECAMOEBIDAE Schaeffer, 1926

Fairly symmetrical, flattened amoebae, with outline oval, elliptical, oblong, flabellate, or semi-circular, moving with a tractor-like rolling of the surface layer, which may have the appearance of a thickened pellicle; discrete pseudopodia rare in locomotion; a variety of nuclear structures and mitotic patterns.

Most adhere poorly to glass; hanging-drop preparations are often unsuccessful and should at any rate be left overnight before being turned over.

1 Pellicle-like surface noticeable by a rather stiff appearance and often by conspicuous wrinkles, either as few parallel, longitudinal, dorsal ridges or numerous and less regularly oriented wrinkles; usually oblong, elliptical, or elongately ovate; in all known freshwater species hyaloplasm as crescent at anterior end, extending posteriorly along sides, but with granuloplasm not set off as thicker, separate mass— 2

— Surface wrinkles slight or absent; hyaline zone thinner than granular mass, either occupying anterior half of amoeba or extending as broad veil around sides— 3

2 Wrinkles, whether or not as regular ridges, conspicuous in most but not all species; normally uninucleate; no cysts or sexual process known— THECAMOEBA

— Shape like *Thecamoeba* (though amoebae do not extend well on slide), but wrinkles often absent or inconspicuous; binucleate, with nuclei always closely apposed; sexual cysts in one species— SAPPINIA

3 (1) No longitudinal wrinkles; outline commonly flabellate, often semi-circular with broad veil around sides, or spatulate with long tail in some; anterior edge commonly smooth but often with shallow temporary indentations in some species; floating form with hyaline radiate pseudopodia tapering markedly, often appearing pointed and sometimes with helical form (Figs 35G, 36C); no cysts known— VANNELLA

— Outline ovate, obovate, circular, semi-circular, elliptical, flabellate, or oblong but seldom if ever spatulate; floating form (Fig. 38D) with a few hyaline pseudopodia not tapering greatly from base, blunt, often bent, and often short— **4**

4 Often slight longitudinal wrinkles near sides, especially when turning; known freshwater species cyst-forming; vesicular nucleus with central nucleolus; nuclear division in known freshwater species with nuclear membrane disintegrating by metaphase— PLATYAMOEBA

— No surface folds or wrinkles, but small bosses formed on surface of hyaline region; vesicular nucleus with multiple peripheral nucleoli; nuclear membrane persisting at least to middle of anaphase— PESSONELLA

Genus THECAMOEBA Fromentel, 1874

A flattened ovoid or oblong (one species very elongate but flattened), with longitudinal surface folds in most species and appearance of thickened 'pellicle' in all; hyaloplasm usually a crescent at anterior end with thin extensions toward posterior end along sides; no discrete pseudopodia or branching except in very elongate forms of one species. Normally uninucleate, with variety of nuclear structures among species.

Although these are considered terrestrial amoebae because often found in soil or mosses, most species have been found also in strictly aquatic habitats, some perhaps only there. No cysts are known, though amoebae of some species, at least, are resistant to desiccation and in dry conditions may assume a disc-like form. These amoebae often prey on smaller amoebae and other protozoa.

The term *endosome* rather than the usual nucleolus is used in the key for aggregations of non-chromosomal intranuclear material, because two named species contain two kinds of intranuclear bodies in each nucleus, differing in appearance and apparently in function.

1 Outline and general surface fairly smooth, with or without several parallel, longitudinal dorsal folds during locomotion; normal locomotive form and activity in fresh wet mounts— 2

— Surface of stationary amoebae moderately to highly wrinkled, becoming somewhat smoother in locomotion but most species with rather wrinkled edges and irregular but usually longitudinal dorsal folds; well extended and active locomotive forms rare on slides— 6

2 With several parallel dorsal folds extending far anteriorly— 3

— Without dorsal folds, or with fold only on boundary of hyaloplasm and granuloplasm in posterior half of amoeba; often a broadly morulate uroid— 5

3 Endosomal material in 2, 3, occasionally more rather large peripheral pieces or band; amoebae 30–80 μm (Fig. 29A–C)—
Thecamoeba striata (Penard, 1890)

— With a single central endosome— 4

4 Length 35–80 μm, occasional giant forms to 170 μm; usually 2–5 dorsal folds; differs from *T. striata* only in nuclear structure and occurrence of gigantism; endosomal texture smooth, not granular (cf. *T. sphaeronucleolus*) (Fig. 29D, E)—
T. quadrilineata (Carter, 1856)
India, Britain, U.S.A.

— Length 30–40 μm; usually about 6 dorsal folds—
T. corrugata Bovee, 1953
U.S.A.

Fig. 29. *Thecamoeba*. A–C, *T. striata* (C, nucleus). D, E, *T. quadrilineata*. (All × 1,000.) *n*, nucleus.

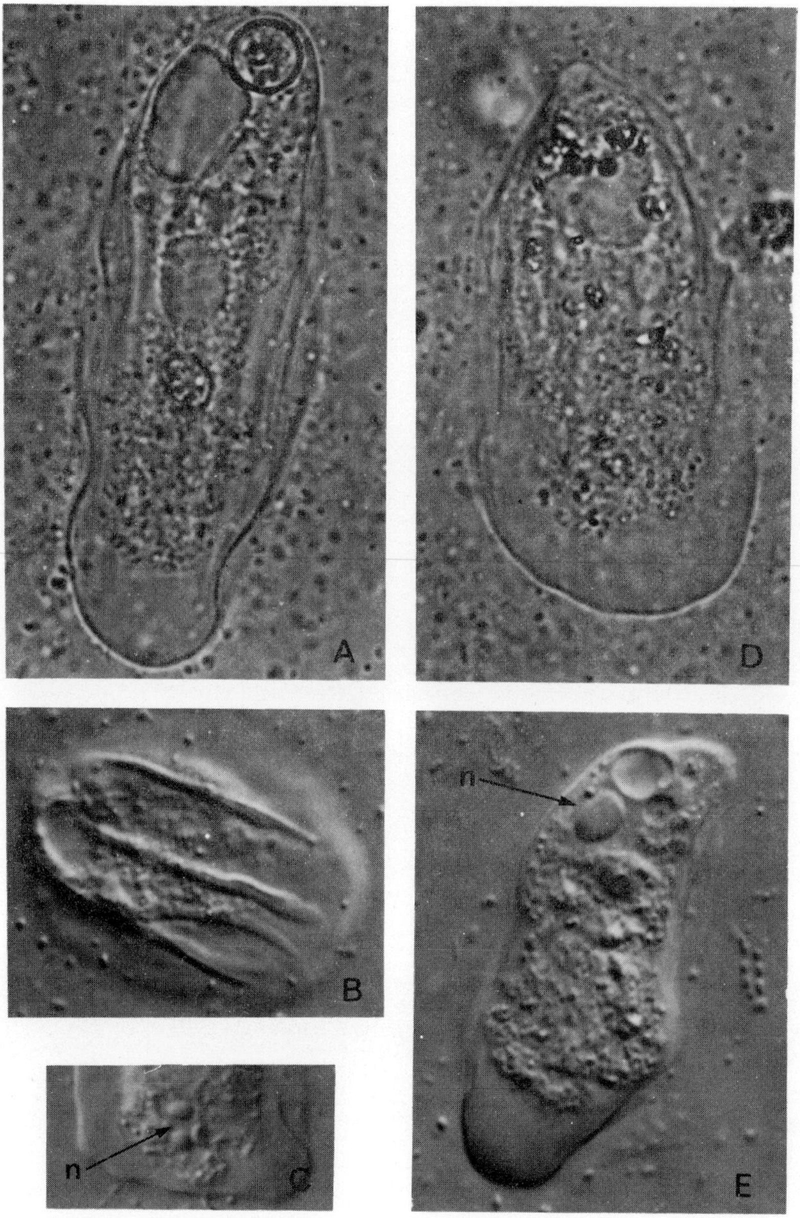
A
D
B
n
C
n
E

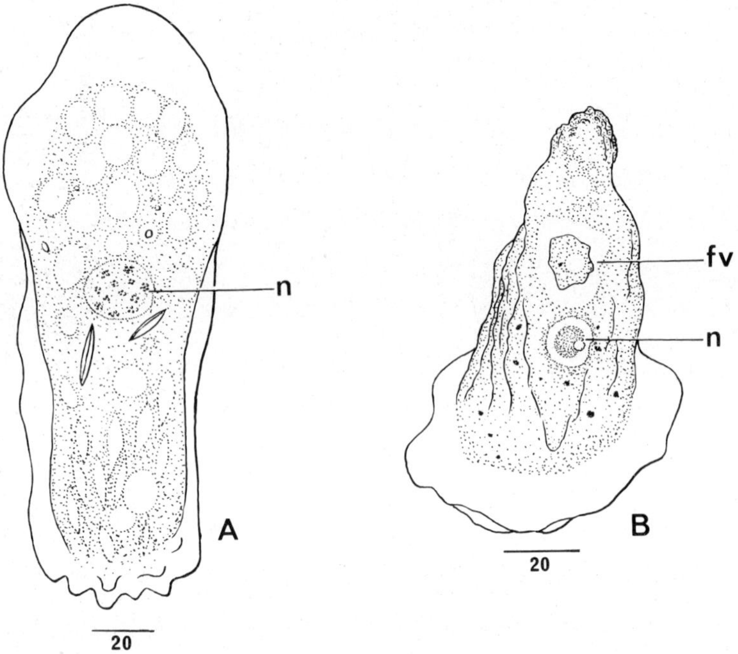

Fig. 30. *Thecamoeba.* A, *T. vesiculata* (after Penard). B, *T. verrucosa* (after Gläser). *fv*, food vacuole; *n*, nucleus.

5 (2) Elongate, typically with L/B *c.* 3 or 4, easily 200 μm; endoplasm may be highly vacuolated; intranuclear material as clumps of fine granules (Fig. 30A)—

T. vesiculata (Penard, 1902)

Europe.

— L/B usually about 2, length 45–85 μm; endoplasm often but not always filled with yellowish, refractile, lipoidal spherules to *c.* 2 μm; two central, approximately spherical, closely apposed endosomal structures (one fainter in appearance than other) in nucleus. (Fig. 31 A–C)— **T. granifera** (Greeff, 1866)

Europe; a similar but smaller species has been observed in both Britain and North America.

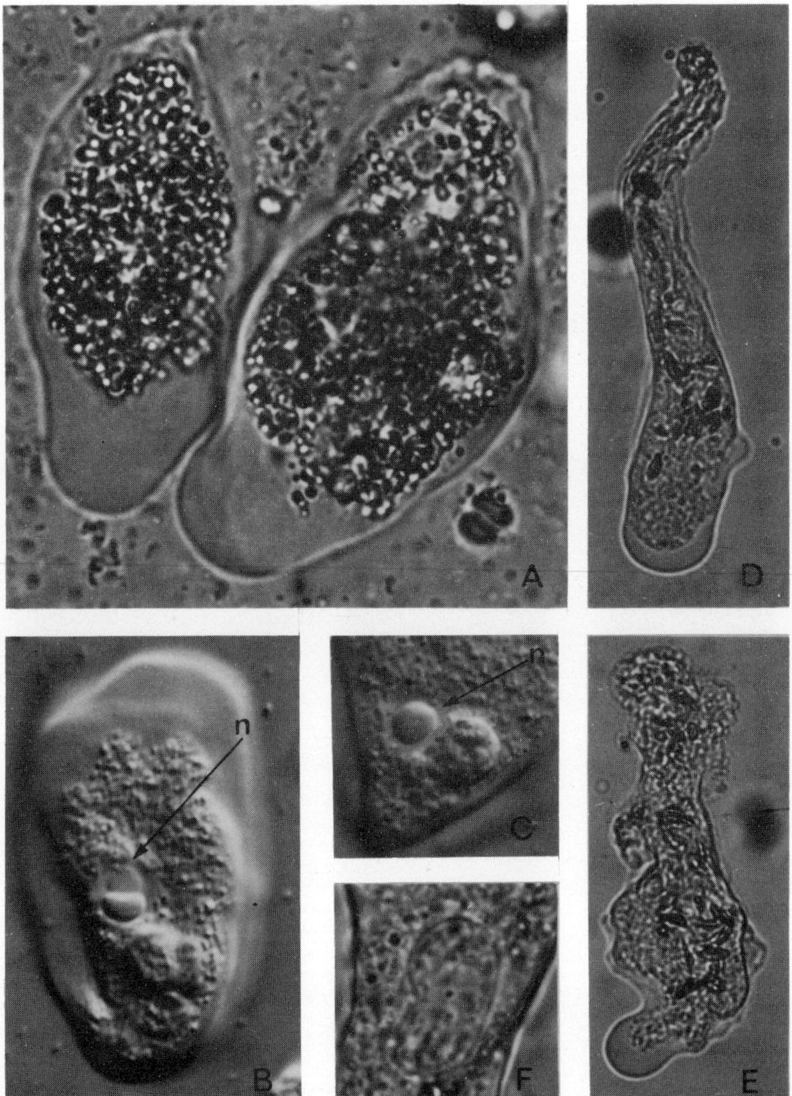

Fig. 31. *Thecamoeba*. A–C, *T. granifera* (only one endosomal body visible in C). D–F, *T. proteoides* (F, nucleus). (A–C, F, ×1,000; D, E, ×300.) *n*, nucleus.

6 (1) Very elongated (L/B to 10·5, mean 4), 60–280 μm, often temporarily branched, giving appearance similar to that of *Amoeba proteus*; more or less oblong nucleus (12–24 μm), with endosomal granules of fairly uniform size beneath nuclear membrane (Fig. 31D–F)— **T. proteoides** Page, 1976

Fresh water; Germany.

— Broader, with L/B less than 3, usually less than 2— **7**

7 Endosomal pieces in layer beneath nuclear membrane— **8**

— Endosomal material not peripherally arranged— **9**

8 Usually 60–200 μm, sometimes larger, usually with wrinkles around periphery in locomotion and sometimes with fine surface wrinkles extending up to three-quarters of body length anteriorly from more or less wrinkled posterior knob; sometimes small hyaline knobs or scallops on anterior periphery; nucleus (15–42 μm) an ellipsoid or prolate spheroid, with endosomal pieces of very varied sizes arranged in layer beneath nuclear membrane (Fig. 32 A–C)— **T. terricola** (Greeff, 1866)

Europe. Distinguish from *T. verrucosa*, with which it is often confused.

— Length 30–80 μm, often broader than most members of genus (L/B 0·8–1·8, mean 1·3); surface moderately wrinkled in stationary amoebae, smoother in locomotion, sometimes with a few dorsal folds converging somewhat posteriorly; nucleus (7·7–13·9 μm) usually ovoidal or ellipsoidal with numerous endosomal pieces in peripheral zone beneath nuclear membrane (Fig. 32D–G)— **T. similis** (Greeff, 1891)

Britain, Germany.

Fig. 32. *Thecamoeba*. A–C, *T. terricola* (C, nucleus). D–G, *T. similis* (F, G, nucleus). (A, ×600; B, ×300; C–E, ×1,000; F, G, ×2,000.) *n*, nucleus.

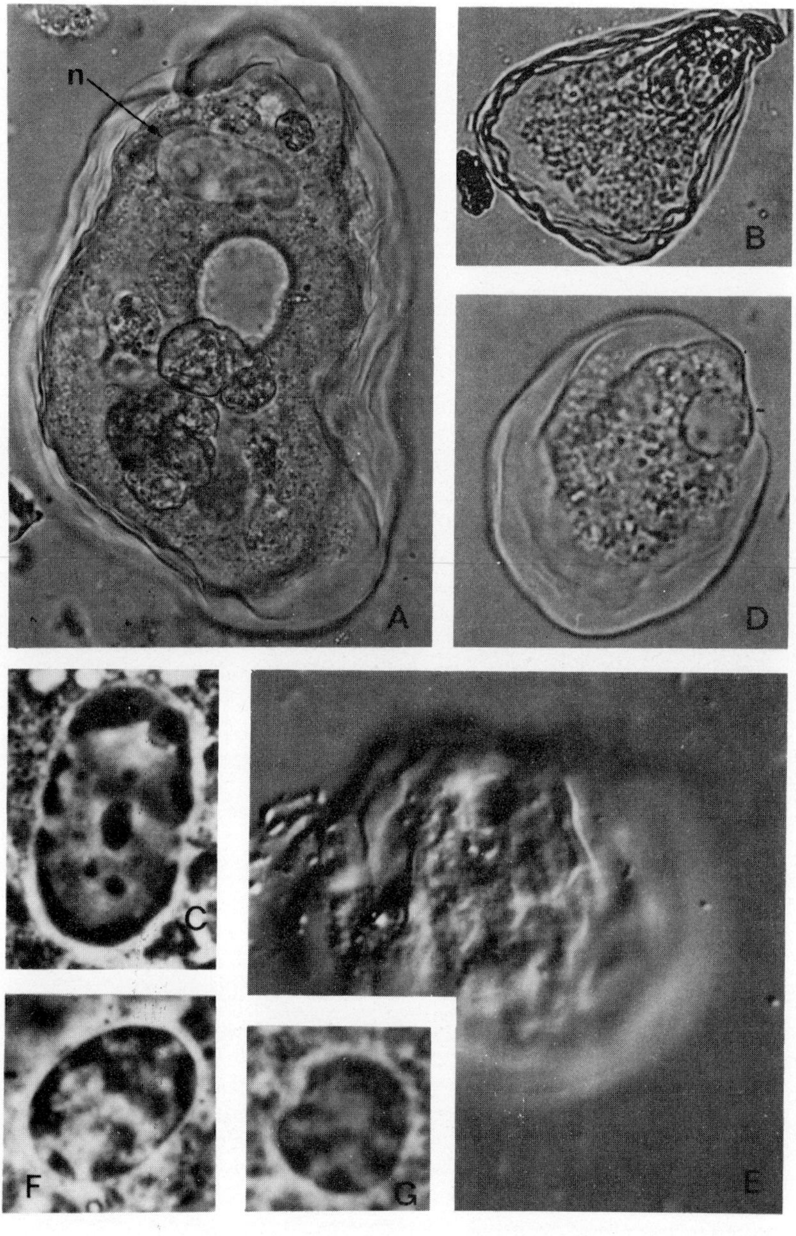
n
A
B
C
D
E
F
G

9 (7) Length 80–130 μm; sometimes with knobby uroid but not strongly tapered posteriorly (not triangular); nucleus spherical or ovoid, endosomal material of coarsely granular texture in rounded or ovoid central mass, sometimes partly or completely divided into 2 or 3 fragments which are not peripherally arranged. (Fig. 33A, B)—

T. sphaeronucleolus (Greeff, 1891)

Europe, U.S.A.

— Length approximately as above or slightly larger, with more tendency to form shrivelled uroid on markedly tapered posterior end (sometimes triangular); two central, closely apposed endosomal structures in nucleus. (Fig. 30B)—

T. verrucosa (Ehrenberg, 1838) sensu Gläser, 1912

Germany.

Fig. 33. A, B, *Thecamoeba sphaeronucleolus* (B, nucleus). C–E, *Sappinia diploidea*. C, locomotive form on agar. D, early sexual cyst, with two gamonts secreting cyst wall. E, mature unicellular, binucleate cyst (methyl green - acetic acid fixation). (A, C, ×1,000; B, D, E, ×2,000.) *n*, nucleus.

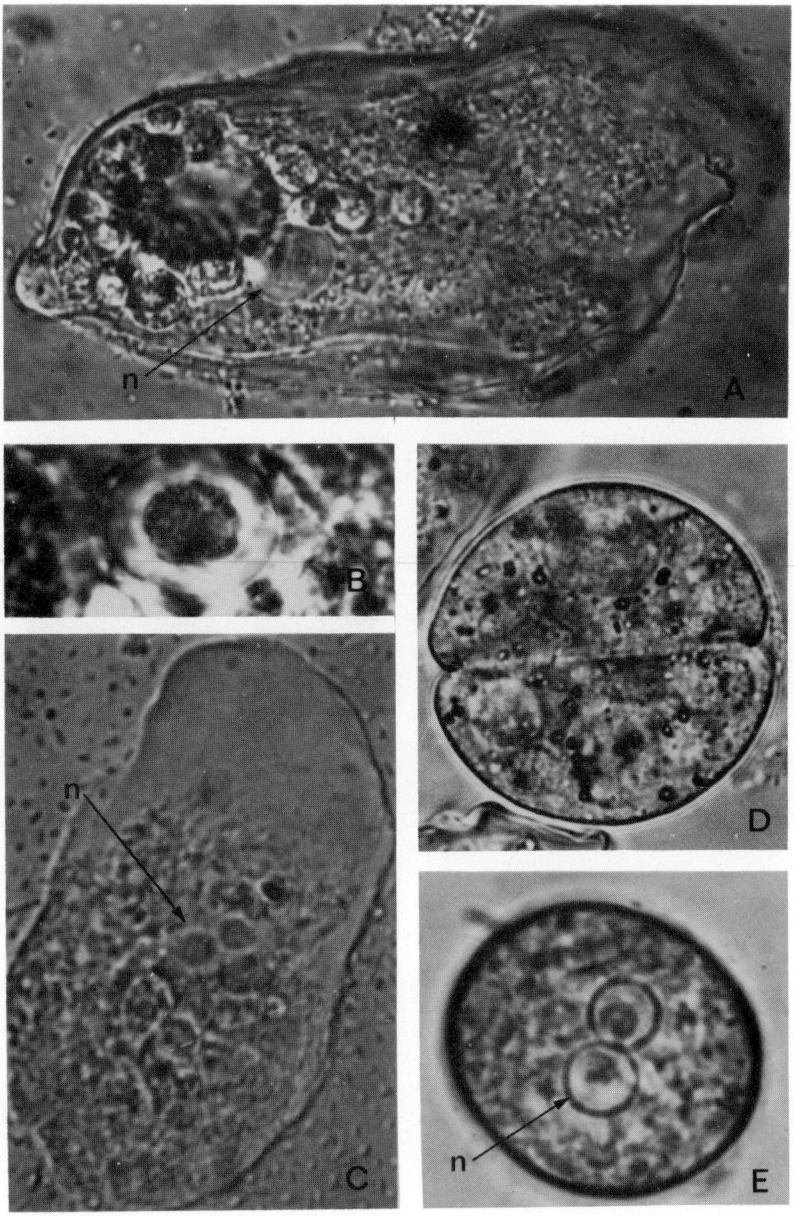
n
A
B
n
C
D
n
E

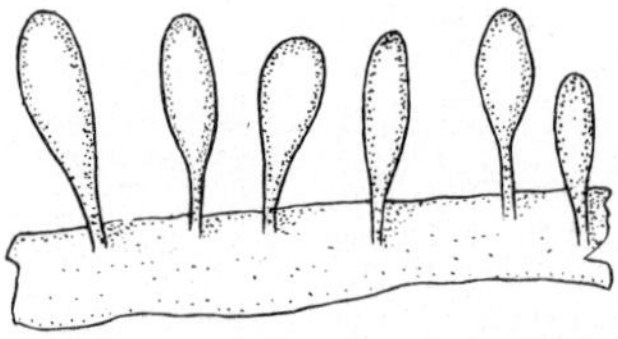

Fig. 34. *Sappinia pedata*, pedicellate 'cysts' (after Dangeard).

Genus SAPPINIA Dangeard, 1896

A flattened ovoid or oblong with hyaloplasm sometimes extending slightly rearward along sides; surface with appearance of pellicle, occasionally wrinkled but without regular ridges; one pair of nuclei, sometimes two or three pairs, the members of a pair closely apposed.

Since these amoebae do not extend well on glass, it is recommended that they be examined on the agar surface. Cysts can be examined on slides.

1 No sexual stage reported; stalked resting stage or cyst (Fig. 34); measurements not reported; amoebae similar to those of next species— **Sappinia pedata** Dangeard, 1896

Soil, horse dung; Europe, North America.

— 43–86 μm; forming sexual cysts which begin as closely apposed pair of binucleate amoebae, the mature cyst unicellular and binucleate, not stalked. (Fig. 33C–E)— **S. diploidea** (Hartmann and Nägler, 1908)

Fresh water, soil, forest litter, mammalian faeces, rectum of lizard; Europe, North America, Egypt, Palestine, West Indies, Japan.

Genus Vannella Bovee, 1965

Flattened, with outline typically flabellate, also spatulate or occasionally oval; hyaline, fan-like, flattened veil occupying anterior one-fourth to one-half and usually extending around sides of thickened posterior granular mass. Usually no pseudopodia or uroidal projections during locomotion; if briefly occurring, these are not filamentous. Floating form commonly with rounded central mass and several slender, hyaline radiate pseudopodia, tapering to narrow tip, with one or more sometimes thickly helical. Single vesiculate nucleus in known species. No cysts known.

This is one of the common genera in fresh water and has even grown out from leaf litter, despite the apparent absence of cysts. There are undoubtedly more species than those in the key, but some descriptions are unsatisfactory. Because species have not often been correctly identified, notes on distribution are omitted.

The genus must be carefully distinguished from *Platyamoeba*. Since observation of the floating form is one means of distinguishing the two genera, it should be noted that the radiate pseudopodia of floating *Vannella* may be short and stumpy for a brief time after the amoebae are suspended in fluid, but then become longer and more pointed; sometimes the long, tapering pseudopodia are seen at once. In *V. simplex* and *V. mira*, at least, the floating form may at times have only one or two pseudopodia incorporating most of the cyto–plasm (comet-shaped). Because of the variation in floating forms within a species, this character does not yet seem of great value for specific identification.

1 Locomotive form in large minority somewhat longer than broad, sometimes spatulate with long posterior extension; greatest dimension of locomotive form 10–25 μm; nucleus 3·4–4·8 μm; settles from floating form in a few minutes; attaches well to glass substratum. (Fig. 35A–D)—
Vannella platypodia (Gläser, 1912)

— Larger than above and with breadth commonly the greatest dimension; seldom an elongated posterior end— 2

2 Breadth usually greater than, sometimes twice as great as length; hyaline veil sometimes only along front, sometimes with ragged edge, rarely and transiently surrounding granular mass completely; occasionally 2, rarely 3 nuclei; greatest dimension of locomotive form 35–80 μm; nucleus 7·7–10·8 μm; slow to settle from floating form; attaches to glass very poorly or not at all. (Fig. 35E–I)—
V. simplex (Wohlfarth-Bottermann, 1960)

— Smaller than above— 3

Fig. 35. *Vannella.* A–D, *V. platypodia,* locomotive (A–C) and floating (D) forms. E–I, *V. simplex,* locomotive (E, F) and floating (G–I) forms. (A, ×400; B, C, E, F, ×1,000; D, G–I, ×250.)

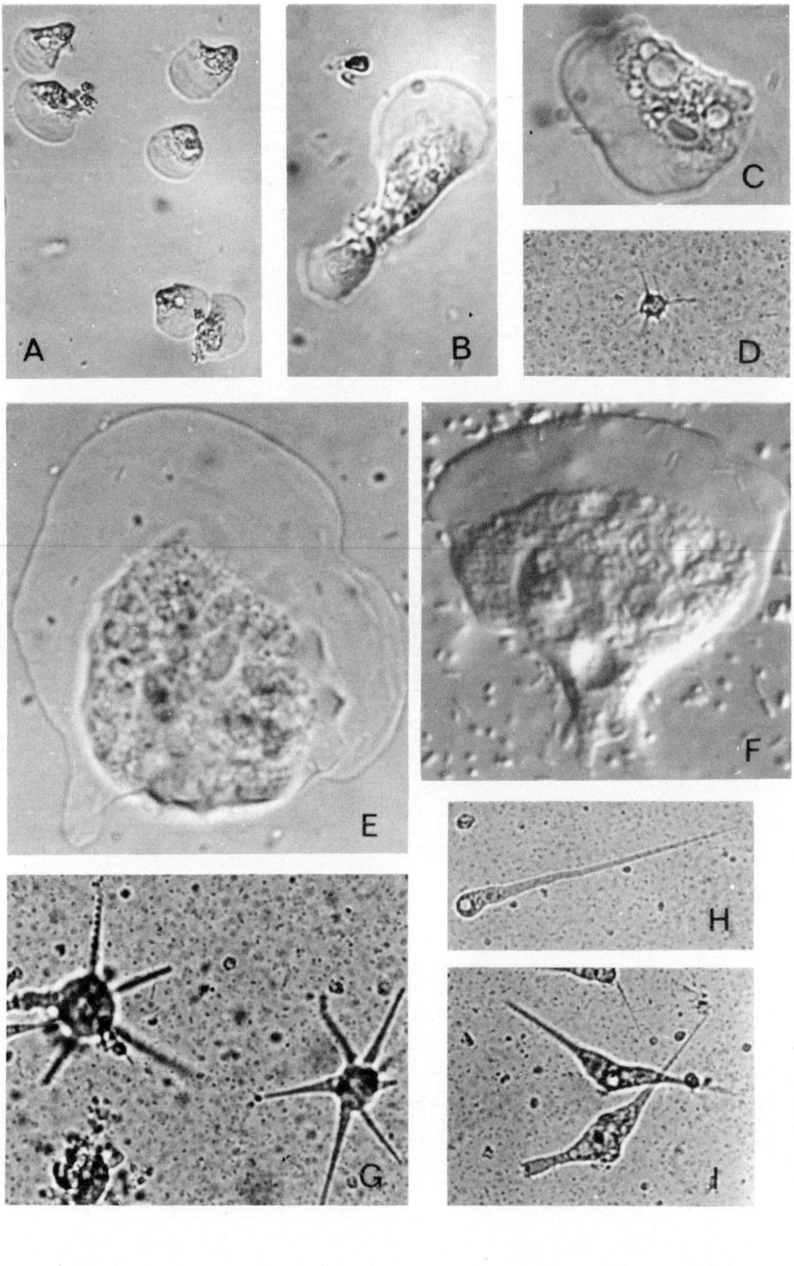
A
B
C
D
E
F
G
H
I

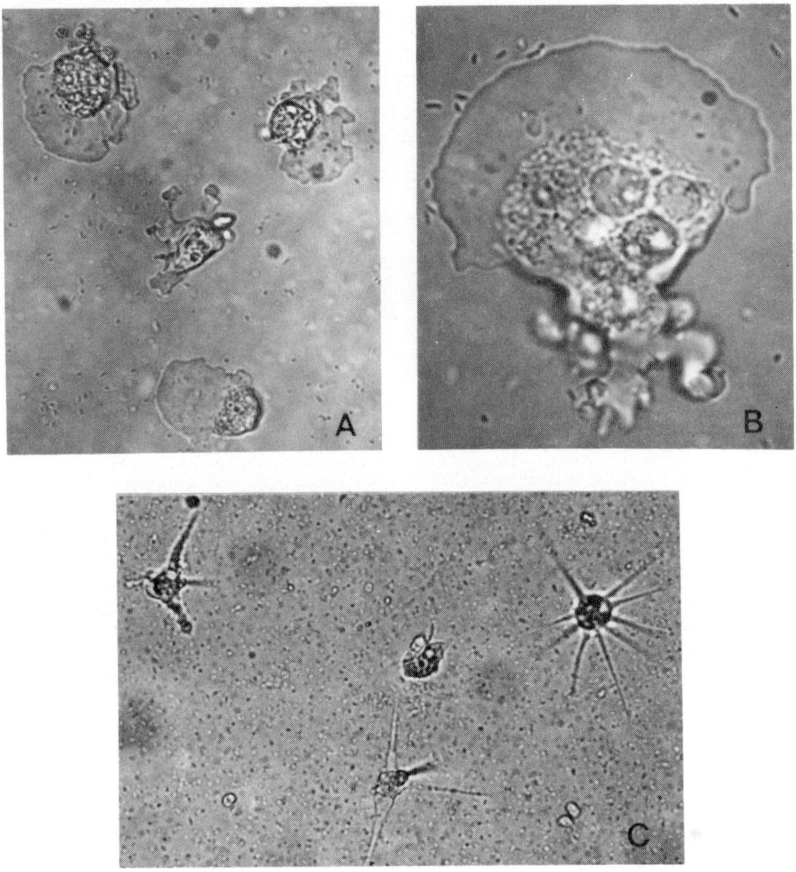

Fig. 36. *Vannella mira*, locomotive (A, B) and floating (C) forms. (A, ×400; B, ×1,000; C, ×250.)

3 Usually flabellate, or somewhat spatulate or oval; greatest dimension of locomotive form 20–55 μm; nucleus 4·8–8·3 μm; slow to settle from floating form; attaches poorly to glass. (Fig. 36)— **V. mira** (Schaeffer, 1926)

Schaeffer described '*Flabellula*' *mira* from salt water, but this freshwater amoeba seems to be morphologically the same.

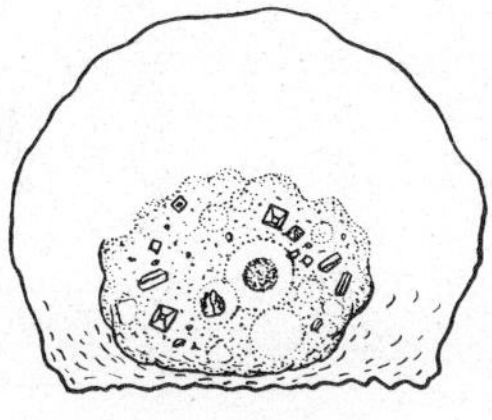

Fig. 37. *Vannella miroides* (after Bovee).

— More or less flabellate, with anterior edge tending to be somewhat wavy; greatest dimension of locomotive form 25–35 μm; nucleus 4–4·5 μm; floating form very regular, with cleanly tapered pseudopodia of almost equal lengths; may be slow to settle from floating form. (Fig. 37)—
V. miroides Bovee, 1965

DURHAM UNIVERSITY
SCIENCE
27 NOV 1976

Genus PLATYAMOEBA Page, 1969

Flattened, with outline oval, elliptical, flabellate, or linguiform; length often greater than breadth; anterior half hyaline, posterior half granular and somewhat thicker; often with folds or wrinkles near and approximately parallel to edge but not definite and constant dorsal ridges. Floating form commonly with blunt, mostly hyaline, radiate pseudopodia. Single vesiculate nucleus which in known freshwater species has a central nucleolus. Mesomitosis accompanied in known freshwater species by early disappearance of nuclear membrane. Known freshwater species cyst-forming.

Only two species have been described from fresh water; the genus appears to be more abundant in salt water. Distinguish carefully from *Vannella*, which is more common in fresh water, and from *Pessonella*.

1 Outline usually oval, often broadly elongated, less often flabellate; anterior edge arched or truncate; cyst spherical or sub-spherical, with two closely apposed smooth wall layers; amoebae 16–35 μm, with L/B 0·6–1·9 (mean 1·2); cyst 7–10·5 μm. (Fig. 38A–E)— **Platyamoeba placida** (Page, 1968)

Fresh water; Britain, U.S.A.

— Often elongated, linguiform, or an elongated oval, with anterior end often truncate; cyst spherical, with smooth inner wall and thinner, often slightly wrinkled but closely applied outer wall; amoebae 15–36 μm, with L/B 1·5–4·5 (mean 2·6); cyst 7–11·5 μm (Fig. 38F–J)— **P. stenopodia** Page, 1969

Fresh water, leaf litter; Britain, U.S.A.

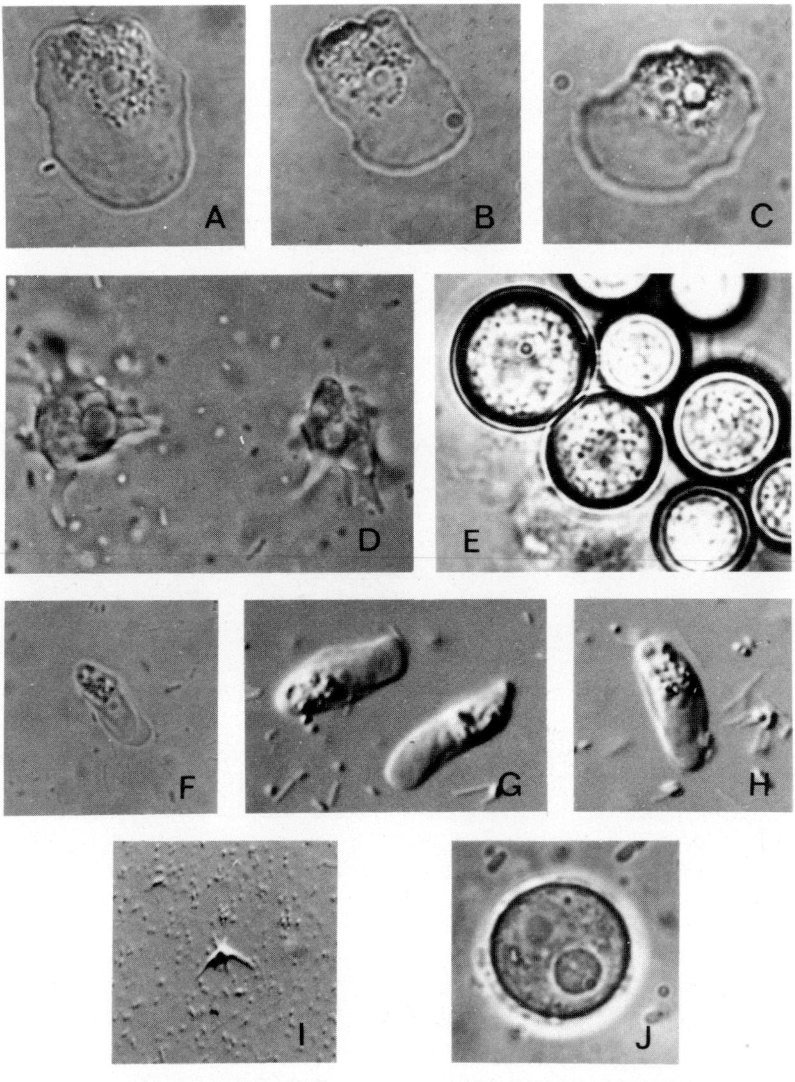

Fig. 38. *Platyamoeba*. A–E, *P. placida*. A–C, locomotive forms. D, floating forms. E, cysts. F–J, *P. stenopodia*. F–H, locomotive forms. I, floating form. J, cyst. (A–D, F–H, ×1,000; E, J, ×2,000; I, ×250.)

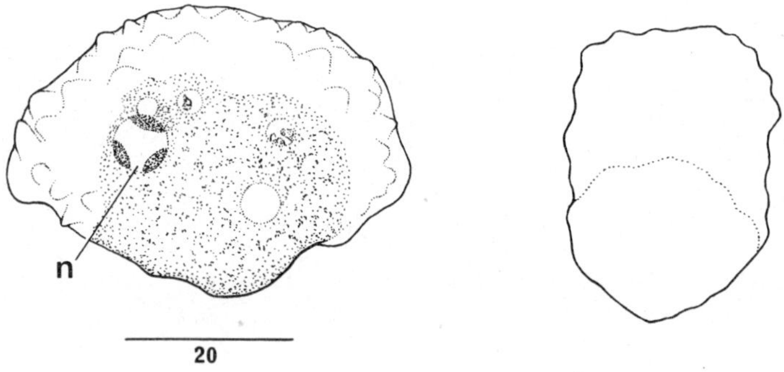

Fig. 39. *Pessonella marginata*, with outline of a second locomotive form (after Pussard). *n*, nucleus.

Genus PESSONELLA Pussard, 1973

Flattened, with outline oval, circular, semi-circular, elliptical or crescentic; breadth usually but not always greater than length; extensive anterior hyaline veil and thicker granular posterior region. No surface folds or wrinkles, but small bosses occurring on surface of hyaline region. No pseudopodia formed in locomotion except small phagocytotic projections which occur especially in posterior region. Floating form with blunt, hyaline, bent radiate pseudopodia. Nucleus with multiple peripheral nucleoli. Nuclear membrane persisting at least to middle of anaphase.

Only one species described. This genus must be distinguished carefully from *Vannella* and especially from *Platyamoeba*.

Approximately 35×45 μm; nucleus 4–8 μm; 2–7 (usually 3–5) nucleolar pieces; no cyst observed. (Fig. 39)—

Pessonella marginata Pussard, 1973

From artificial fertilizer made with straw; France.

Family HYALODISCIDAE Poche, 1913

Fairly regularly discoid or somewhat fan-shaped, with breadth usually greater than length, with central, post-central, or posterior granular hump surrounded completely or anteriorly and laterally by flattened hyaline border; sometimes with narrowly conical hyaline projections or pseudopodia; rolling movement often apparently involving entire cell; no tectum of scales.

1 Conical sub-pseudopodia often absent; flattened hyaline border often surrounding granular hump completely— HYALODISCUS

— Short conical sub-pseudopodia always present, often in clusters of two or three; hyaline border seldom continuing behind granular hump— FLAMELLA

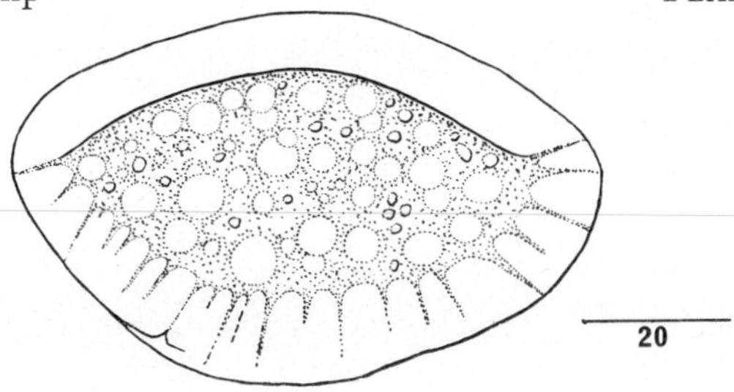

Fig. 40. *Hyalodiscus rubicundus* (after Cash).

Genus HYALODISCUS Hertwig and Lesser, 1874

Discoid, with breadth the greatest dimension, with central or post-central granular hump surrounded by a flattened hyaline border; sometimes fine, short, conical projections from hyaloplasm.

Only one certain species from fresh water; apparently not common.

Distinguish carefully from *Cochliopodium*.

Granuloplasm containing red, greenish-brown and reddish-brown inclusions, apparently remains of ingested plant material; occasional folds of hyaline zone producing appearance of veins of granuloplasm; breadth 30–80 μm. (Fig. 40)—
Hyalodiscus rubicundus Hertwig and Lesser, 1874

Fresh water; Europe.

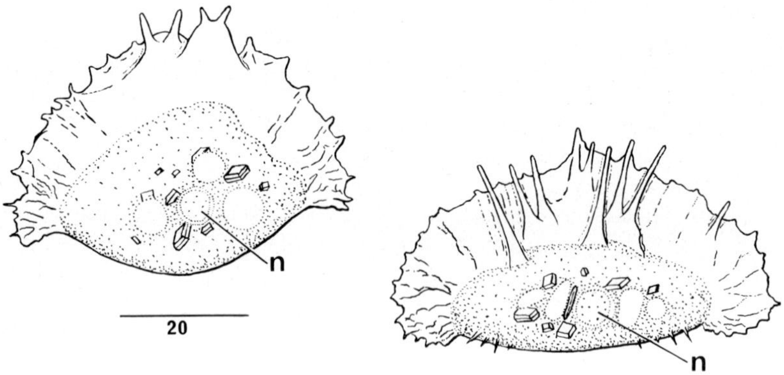

Fig. 41. *Flamella citrensis* (after Bovee; both to same scale). *n*, nucleus.

Genus FLAMELLA Schaeffer, 1926

A flattened ovoid or fan shape with posterior granular mass, extensive flattened anterior and lateral hyaline zone, and numerous slender, conical, hyaline, usually short pseudopodia fusing with hyaline zone; locomotion and change of shape rapid; nucleus obscure.

One known freshwater species. On the basis of locomotive form alone, it would seem that *Hartmannella klitzkei* Arndt, 1924, might belong to this genus.

Pseudopodia produced in 2's and 3's from pyramidal bases at anterior edge of granuloplasm; both pseudopodia and bases merging with hyaline zone; occasionally a few short, trailing filaments; breadth increasing with rate of locomotion, 35–55 μm; floating form spherical with hyaline papillae; anaerobic. (Fig. 41)— **Flamella citrensis** Bovee, 1956

From citrus-pulp wastes in freshwater tanks; U.S.A.

Family ECHINAMOEBIDAE Page, 1975

Amoebae flattened or temporarily elongate, producing filose pseudopodia or micro-spines **from hyaline region**; outline often highly irregular, sometimes spatulate or flabellate; no flagellate stages; cyst walls, when cysts formed, without actual or potential opercula (see p. 29); no spores, sorocarps, or fruiting bodies.

This recently established family is intended to include all Amoebida with truly filose, hyaline pseudopodia. They must be distinguished from amoeboid organisms with fine granular or anastomosing pseudopodia, and from the Acanthamoebidae, in which the pseudopodia are often so fine as to merit the descriptive term filose but may also be much coarser. They must be distinguished also from organisms with myxamoebae which produce micro-spines and also have plasmodial or pseudoplasmodial stages or sorocarps (myxomycetes and acrasiomycetes).

1 Many filose pseudopodia, several micrometres long, produced from one or more hyaline zones; amoebae often very flattened, often spatulate— FILAMOEBA

— Short, spine-like projections (micro-spines) from hyaline zone— **2**

2 Usually flattened, often somewhat triangular, sometimes elongate; extensive anterior hyaline zone; spines sometimes absent from more elongate forms in rapid locomotion; no eruptive hemispherical pseudopodia; a single central nucleolus; cysts spherical or ovoid with outer wall smooth— ECHINAMOEBA

— Usually flattened, expanded irregular shape; hyaline zone a rather narrow band across anterior end or extending around sides of granuloplasm in broad forms; sometimes a limax form with eruptive hemispherical pseudopodia (resembling vahlkampfiid amoebae); one or more peripheral nucleoli in single described species; outer cyst wall of that species rippled— STACHYAMOEBA

Genus Filamoeba Page, 1967

Clear, filose, sometimes furcate but never anastomosing subpseudopodia produced from one or more hyaline zones, the length of the filopodia usually greater than the depth of the hyaloplasmic border; mesomitosis.

Only one described species. Distinguish carefully from filose genera not belonging to this subclass.

Outline varying from semi-circular to spatulate, sometimes multipolar with two or more hyaline zones; sometimes flattened into a thin sheet; up to 15 or more contractile vacuoles; locomotive and relatively inactive but extended forms 13–48 μm; cyst smooth, rather thick-walled with outline circular, irregularly ovate, or reniform, with no pores, 8–15 μm. (Fig. 42A–E)— **Filamoeba nolandi** Page, 1967

Fresh water; U.S.A., Czechoslovakia.

Genus Echinamoeba Page, 1975

Often somewhat triangular with anterior end broad, sometimes elongate; micro-spines usually produced from extensive anterior hyaline zone but sometimes temporarily absent in vigorous locomotion; locomotion of more elongate forms not explosive. Cysts smooth, spherical or ovoid.

1 Expanded and fan-like to elongate and limax-like, with irregular anterior edge; cysts with outer wall closely apposed to inner, occasionally raised at some points, with a few coarse cytoplasmic granules; amoebae 7–22 μm, nucleus 2·1–2·8 μm, with small nucleolus, *c.* 0·5 μm; cysts 3·8–8·4 μm. (Fig. 42F–H)— **Echinamoeba exundans** (Page, 1967)

Fresh water; U.S.A., Czechoslovakia.

Fig. 42. A–E, *Filamoeba nolandi*. A–C, locomotive forms. D–E, cysts. F–H, *Echinamoeba exundans*. F, G, locomotive forms. H, cysts. (A–C, F, G, ×1,000; D, E, H, ×2,000.)

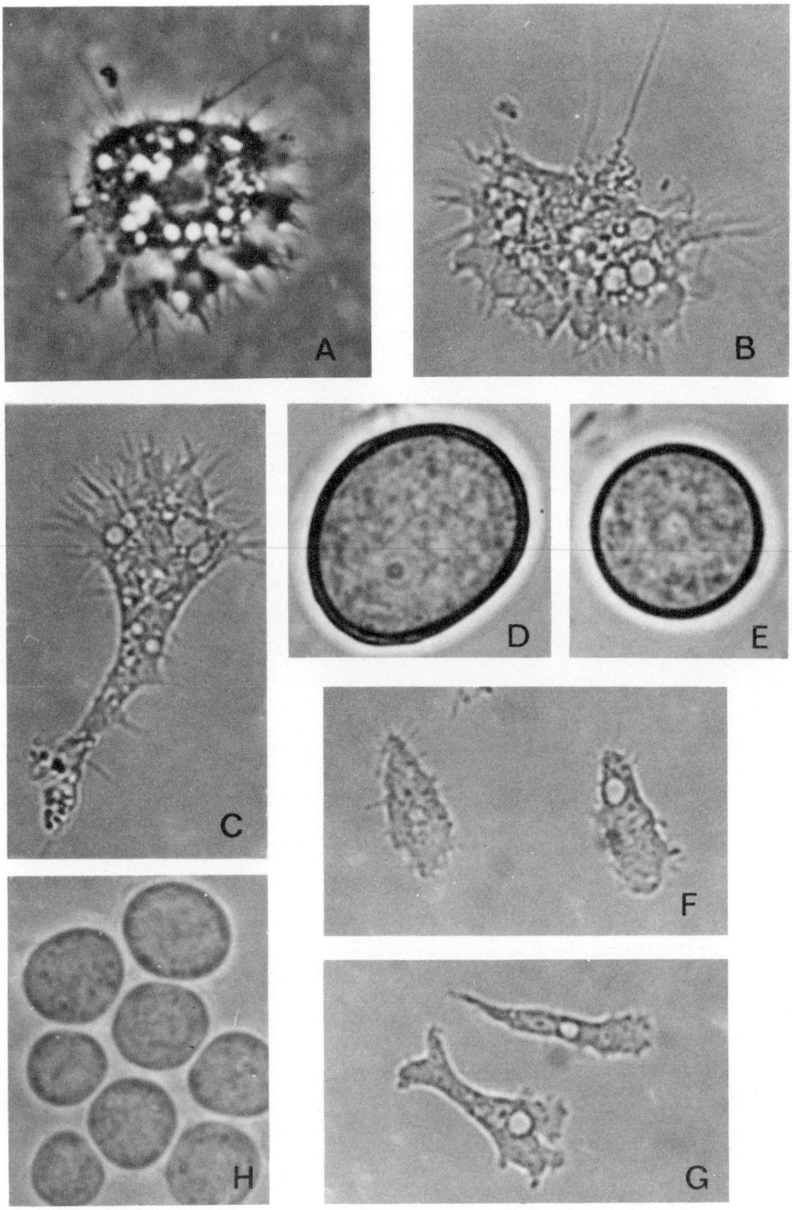
A
B
C
D
E
F
G
H

— Tending to have triangular outline, seldom if ever broadened or spatulate; cysts with narrow space between inner and outer walls and without noticeable coarse cytoplasmic granules; amoebae 6·5–21 μm, nucleus 2–2·8 μm, with nucleolus 1–1·5 μm; cysts 5·6–10 μm. (Fig. 43A–D)—

Echinamoeba silvestris Page, 1975

Leaf litter; Britain.

Genus Stachyamoeba Page, 1975

Usually rather thickly flattened, with expanded irregular shape; micro-spines often produced from rather narrow anterior or antero-lateral hyaline band; sometimes a limax form with eruptive, hemi-spherical pseudopodia; one or more peripheral nucleoli in the one known species; cysts of that species with rippled outer wall.

Irregular outline occasionally becoming spatulate or somewhat elongate; sometimes taking true limax form with eruptive hyaline bulges and no micro-spines; usually one prominent lipid droplet and smaller ones; cysts spherical, sometimes reniform or crescentic, with outer wall raised into many more or less acute peaks; amoebae 15–33 μm, nucleus 2·8–4·6 μm, cysts 8·4–13 μm. (Fig. 43E–L)—

Stachyamoeba lipophora Page, 1975

Soil; Scotland.

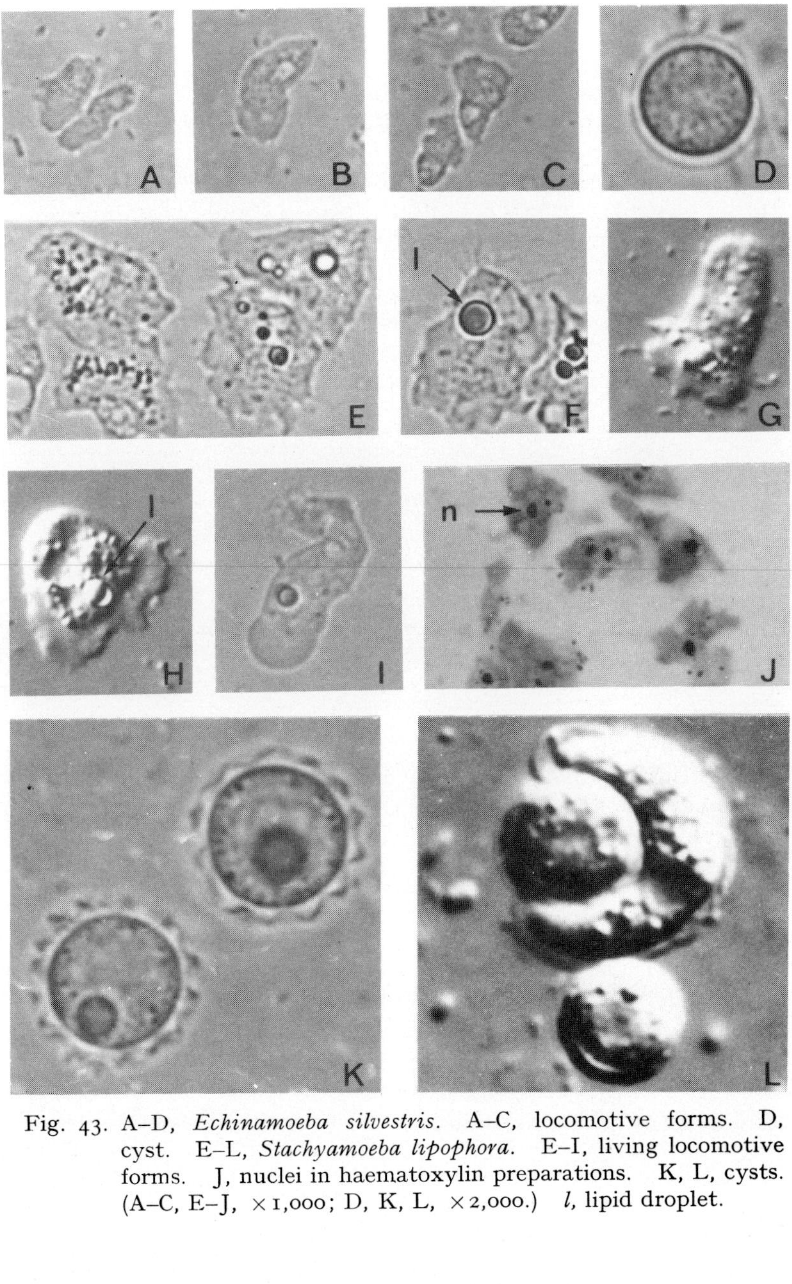

Fig. 43. A–D, *Echinamoeba silvestris*. A–C, locomotive forms. D, cyst. E–L, *Stachyamoeba lipophora*. E–I, living locomotive forms. J, nuclei in haematoxylin preparations. K, L, cysts. (A–C, E–J, ×1,000; D, K, L, ×2,000.) *l*, lipid droplet.

Family ACANTHAMOEBIDAE Sawyer & Griffin, 1975

A single known genus, as the family is understood here. Characters of the family are those given for the genus *Acanthamoeba*.

Genus Acanthamoeba Volkonsky, 1931

Broad hyaline zone from which are produced several to many slender, tapering, flexible, sometimes furcate projections (acanthopodia), which may appear rounded or pointed at tip; outline oval, elongate, or irregular; extranuclear centrospheres reported; cysts polyhedral or thickly biconvex, with cellulose-containing wall consisting of more or less polygonal or stellate endocyst and more or less rippled ectocyst; excystment by removal of operculum (see p. 29) at point of contact between endocyst and ectocyst.

Although the genus is well characterized, especially by the cyst, the status of species is still not well clarified. Some workers consider *A. astronyxis* and *A. comandoni* synonymous, and some consider *A. palestinensis* to be a junior synonym of *A. castellanii* because of biochemical similarities. In this key, *A. rhysodes* (Singh, 1952) is considered a synonym of *A. castellanii*. The species *A. griffini*, though described from a marine source, is included because it multiplies and encysts well on freshwater agar media.

Acanthamoeba is very common in soil and fresh water and has also been isolated from sources within laboratories. Geographical sources are not given because some of these species seem nearly universal and because it is often not possible to judge the accuracy of identifications.

There are indications that strains of some species not designated in the key as pathogenic are at least mildly pathogenic. Until recently, tests of pathogenicity usually were for invasiveness of the central nervous system by nasal instillation. Singh & Das (1970) found *A. rhysodes* pathogenic to mice, though less so than *A. culbertsoni*, which apparently must be considered a pathogenic species. Some strains of *Acanthamoeba* have been cleared of pathogenicity via nasal instillation. Recently, however, strains of *A. polyphaga* and *A. castellanii* have been isolated from human corneal ulcers, and in at least some cases the patient has lost the sight of the affected eye (Nagington *et al.*, 1974). These ocular

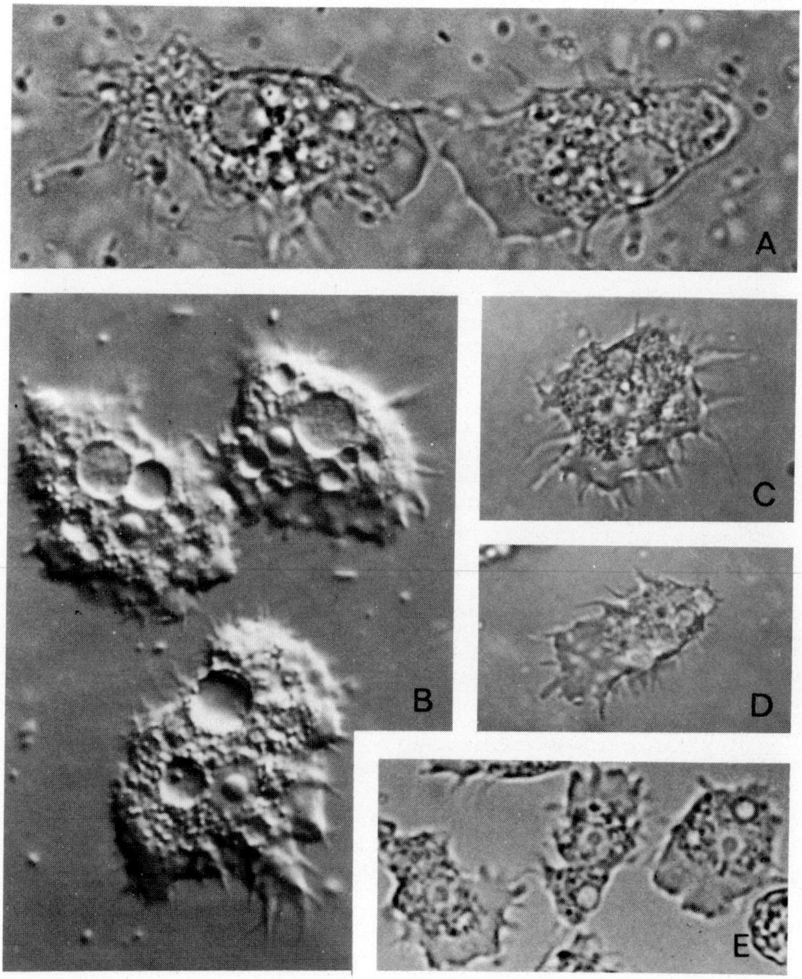

Fig. 44. *Acanthamoeba*, locomotive forms. A, *A. astronyxis*. B–D, *A. castellanii*. E, *A. polyphaga*. (All $\times$ 1,000.)

strains grow better at 23°C than at 37°C. See suggestions on handling of suspected pathogens on page 20.

The key is based on cysts, which are easily obtainable, and the measurements given are cyst diameters.

1 Ectocyst circular or nearly so, gently rippled or smooth; endocyst usually stellate, with tips of rays approximately at general level of ectocyst, not in depressions— **2**

— Endocysts not predominantly stellate or, if endocysts predominantly stellate, with ectocyst closely following endocyst contour— 3

2 All rays of endocyst usually contact ectocyst in approximately same plane; 14–32 μm (mean 19·2–22 μm). (Figs 44A, 45A)— **Acanthamoeba astronyxis** (Ray & Hayes, 1954)

— Rays of endocyst contact ectocyst in several planes; 15–35 μm (mean 20–25·6 μm)— **A. comandoni** Pussard, 1964

3 (1) Endocyst nearly rounded or with only slight angles— **4**

— Endocyst usually with pronounced polygonal or stellate shape— **5**

4 Ectocyst closely apposed to endocyst and no more than slightly rippled; grows well at 23°C; 9–21 μm (mean 14–16·9 μm). (Fig. 45B)— **A. palestinensis** (Reich, 1933)

— Ectocyst more noticeably wrinkled; grows poorly at 23°C, well at 37°C; pathogenic to mice by nasal instillation; 11·6–18 μm (mean 14·9 μm). (Fig. 45C, D)— **A. culbertsoni** (Singh & Das, 1970)

Fig. 45. *Acanthamoeba*, cysts. A, *A. astronyxis*. B, *A. palestinensis*. C, D, *A. culbertsoni* (two strains). E–G, *A. griffini*, from culture on agar made with amoeba saline. (All ×1,200.)

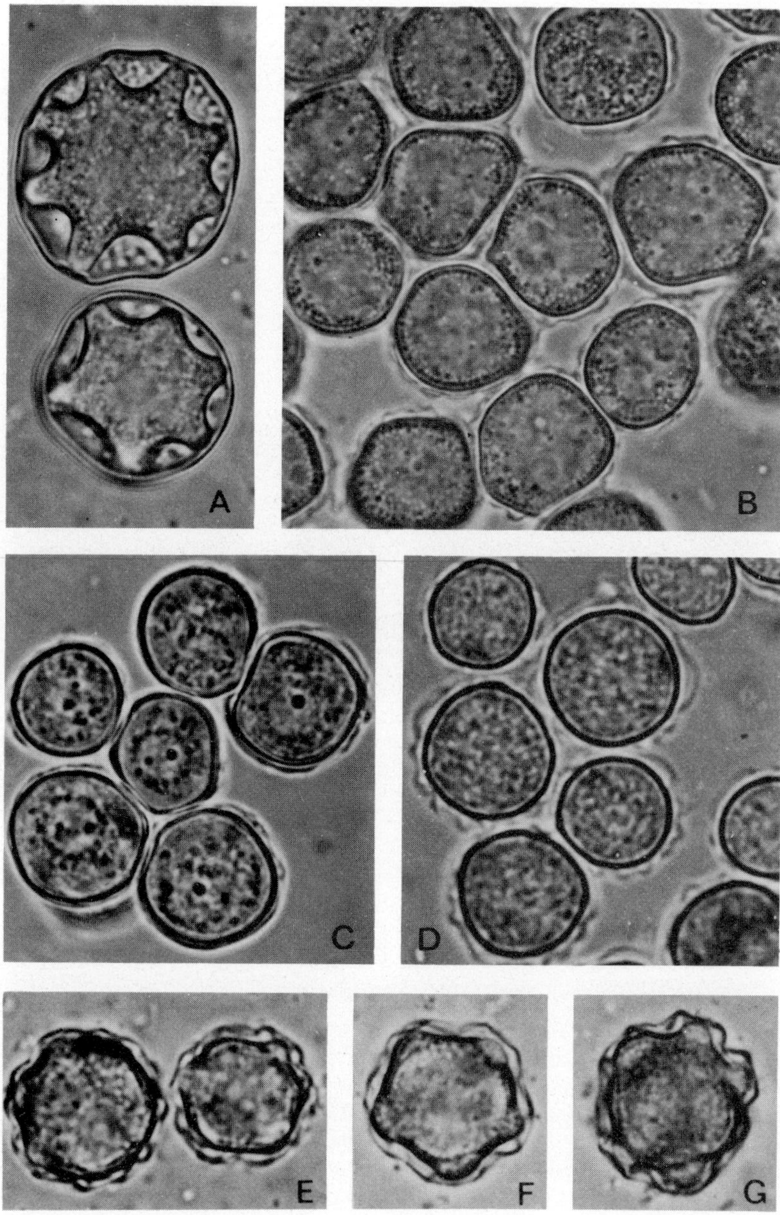
A
B
C
D
E
F
G

5 (3) Over-all cyst shape, with two walls closely apposed, cog-like on distilled-water agar, stellate with blunt rays on seawater agar of salinity 24–28 $^{0}/_{00}$, 14–25 μm on distilled-water agar, 13–17 μm on seawater agar (Fig. 45E–G)—
A. griffini Sawyer, 1971

— Endocysts usually polygonal, occasionally stellate; tip of endocyst ray usually in depression of wrinkled or rippled ectocyst— **6**

6 Endocyst polygonal or, if stellate, usually with six or more truncate rays in one plane; ectocyst usually loosely applied to endocyst, rippled or wrinkled; 8–29 μm (mean 14–20 μm). (Figs 44B–D; 46A–C)— **A. castellanii** (Douglas, 1930)

— Endocyst polygonal, often pentagonal or rectangular, sometimes triangular in outline, rarely stellate; if stellate, rarely more than five or six rays in one plane and with only shallow concavity between rays; ectocyst loosely or very closely applied, often but not always conforming to endocyst shape more than does ectocyst of *A. castellanii*; often wrinkled; 8–21 μm (mean 12–15 μm). (Figs 44E, 46D–F)—
A. polyphaga (Puschkarew, 1913)

Possibly the most common amoeba in many soils. Some strains described before 1913 from endozoic sources may belong to this species but cannot be definitely identified.

Fig. 46. *Acanthamoeba*, cysts. A–C, *A. castellanii* (three strains). D–F, *A. polyphaga* (two clonal strains and one culture of field material). (All × 1,200.)

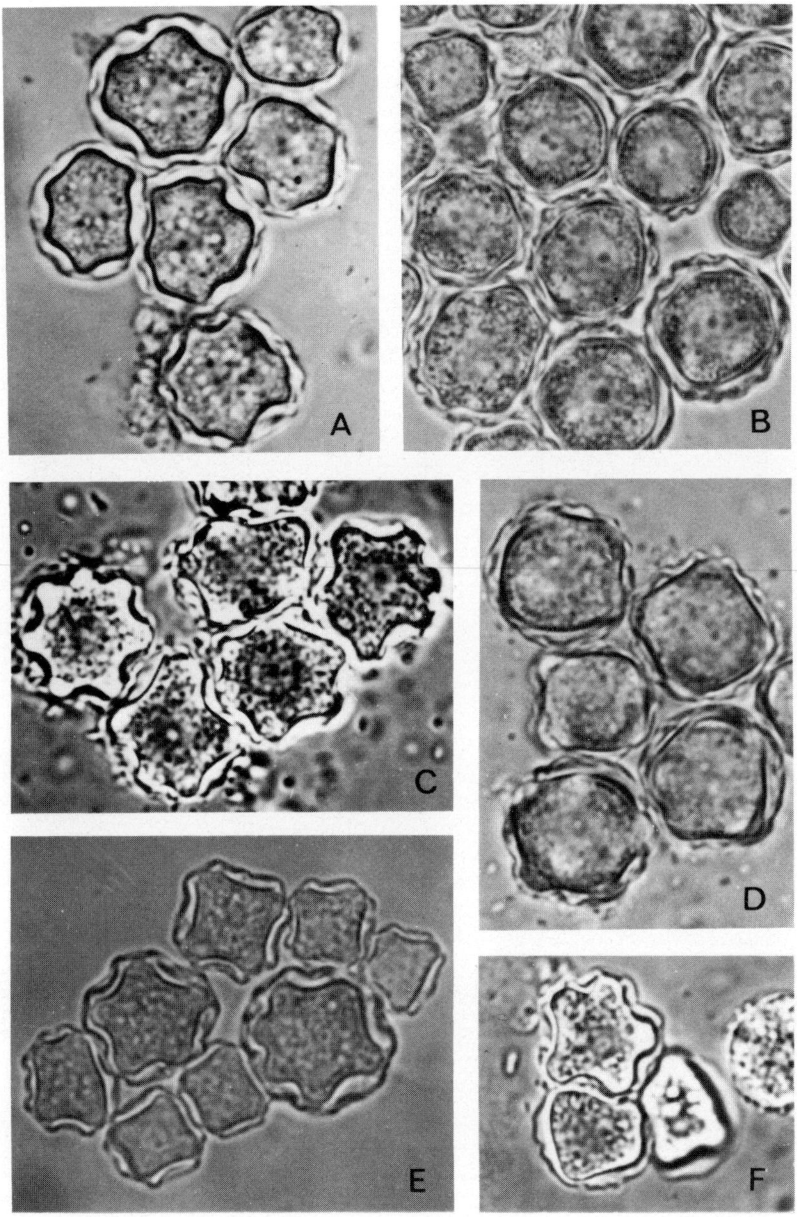
A
B
C
D
E
F

Family PARAMOEBIDAE Poche, 1913; emend.

Conical but not filose sub-pseudopodia produced from anterior hyaline zone, sometimes elsewhere; length almost always greater than breadth, not discoid; cysts not produced by most species, and when formed, excystment not by removal of operculum.

This is the family better known by the junior synonym Mayorellidae. No member of the type genus *Paramoeba* has been reported from habitats other than salt water, and no genus except *Paramoeba* has a DNA-containing Nebenkörper.

1 Pseudopodia of any one amoeba of similar lengths— 2

— Some pseudopodia more or less short cones, others much longer and sometimes waving— 3

2 Pseudopodia relatively short digitiform cones or mamilliform pyramids, with bluntly rounded tip, rarely or never furcate, seldom if ever waving, usually few; amoebae usually rather elongated, with anterior end broader and posterior end usually somewhat bulbous— MAYORELLA

— Pseudopodia tapering to somewhat pointed tip, sometimes furcate, sometimes bending, prominent along sides as well as at anterior end; cell often irregularly ovoid or pyriform in locomotion— DINAMOEBA

3 (1) With short, rather narrowly conical pseudopodia and often one or more long, rod-like hyaline pseudopodia, which may wave somewhat on retraction; over-all appearance spiny because of pseudopodia carried back along sides— VEXILLIFERA

— Producing short, conical pseudopodia and often, from the dorsal surface, a long one which may wave somewhat and dip to the substratum, after which the amoeba flows through it— 4

4 Shape in rapid locomotion awl-like, with posterior end the broadest region at that stage; nucleus with several nucleolar pieces clustered in centre or dispersed— SUBULAMOEBA

— Triangular to clavate in locomotion, nucleus with a single central nucleolus— OSCILLOSIGNUM

Genus MAYORELLA Schaeffer, 1926

Digitiform or mamilliform conical, blunt, hyaline sub-pseudopodia of similar lengths, rarely or never furcate, seldom if ever waving; sub-pseudopodia usually few, in some species overtaken by hyaline wave and obliterated in rapid locomotion; amoebae usually fairly long and rather flattened; posterior end usually rather bulbous, often with morulate uroid; covering of complex scales (*c.* 0·5 μm) in some but not all species, not distinguishable with light microscope.

1 With zoochlorellae; subpseudopodia mamilliform pyramids, often obliterated by advancing hyaline region; 95–160 μm. (Fig. 51A, B)— **Mayorella viridis** (Leidy, 1874)
Fresh water; U.S.A., Britain.

— Without zoochlorellae— **2**

2 Sub-pseudopodia digitiform cones, seldom obliterated in rapid locomotion in most species, in some with a web-like hyaline sheet between adjacent pseudopodia; floating forms of most species with long, tapering, radiate pseudopodia— **3**

— Sub-pseudopodia broadly based mamilliform or nearly hemispherical pyramids, in some species often obliterated by advancing hyaline region; floating forms usually without long radiate pseudopodia— **14**

3 Maximum length more than 60 μm— **4**

— Maximum length 60 μm or less— **8**

4 Elongate, with L/B often 3 or more, usually with only one to several pairs of pseudopodia, without prominent ridges extending posteriorly from pseudopodia— **5**

— Broader, with fairly numerous conical sub-pseudopodia, often prominent longitudinal dorsal ridges— **7**

5 Length 60–130 μm, L/B to 3 or slightly more, with approximately 4 narrowly conical pseudopodia on pyramidal bases, not connected by hyaline web; floating form with slender radiate pseudopodia; nucleus 8 μm. (Fig. 47A, B)—
M. lacona Bovee, 1970

Sewage filters; U.S.A.

— Elongated form in rapid locomotion more limax-like, with anterior tip rounded and any pseudopodia either beneath it or posterior to it; conical pseudopodia of broader locomotive form connected by hyaline web— **6**

6 Length 60–100 μm in moderate to rapid locomotion, L/B to 4; sub-pseudopodium on either side just behind anterior end of limax-like form; uroid morulate or wrinkled; floating form radiate; nucleus 8–9 μm. (Fig. 47C, D)—
M. limacis Bovee, 1970

Fresh water; U.S.A.

— Length 90–140 μm in moderate to rapid locomotion (shorter in slower locomotion), L/B to 5·6; pseudopodia of limax-like form produced ventral to anterior end, difficult to see; uroidal knob with several short, conical filaments; floating form irregularly rounded up, not radiate; nucleus 8 μm. (Fig. 47E–G)—
M. clavella Bovee, 1951

Fresh water; U.S.A.
(Distinguish the preceding two species carefully from *M. clavabellans*, which also has a limax-like form.)

Fig. 47. *Mayorella*. A, B, *M. lacona*. C, D, *M. limacis*. E–G, *M. clavella*. (All after Bovee.)

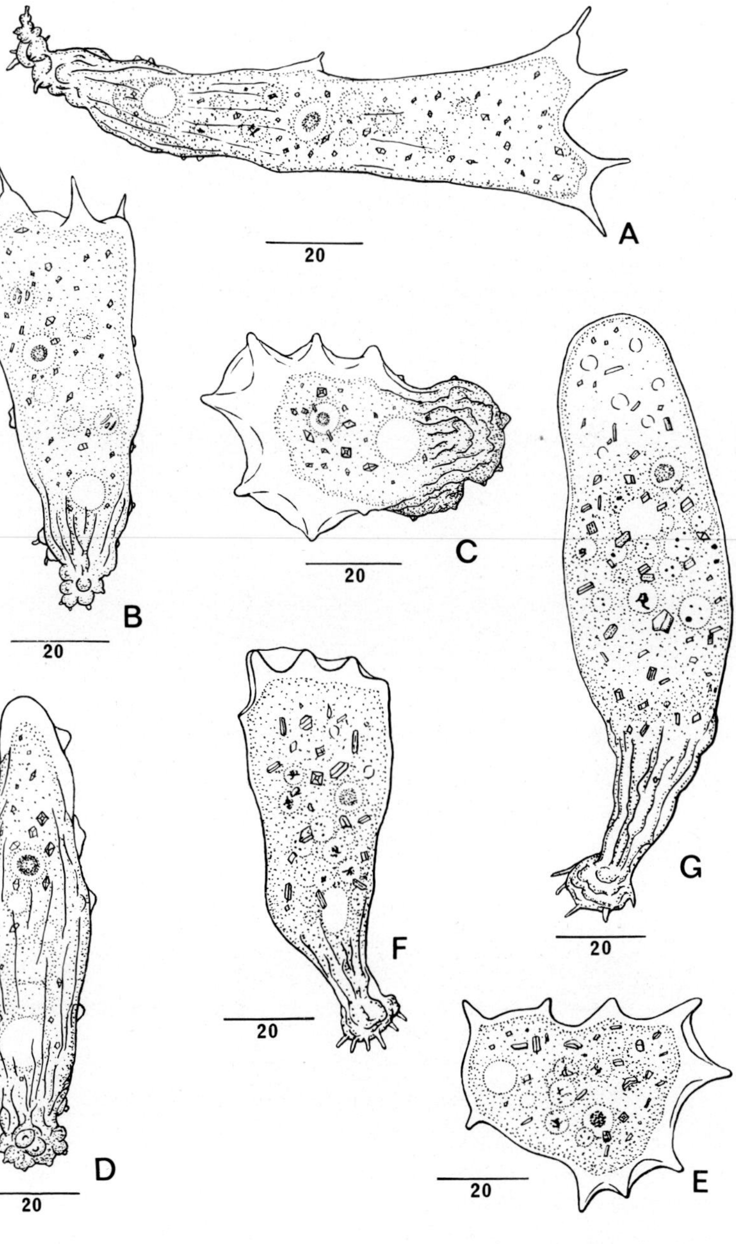
A
20
B
20
C
20
G
20
D
20
F
20
E
20

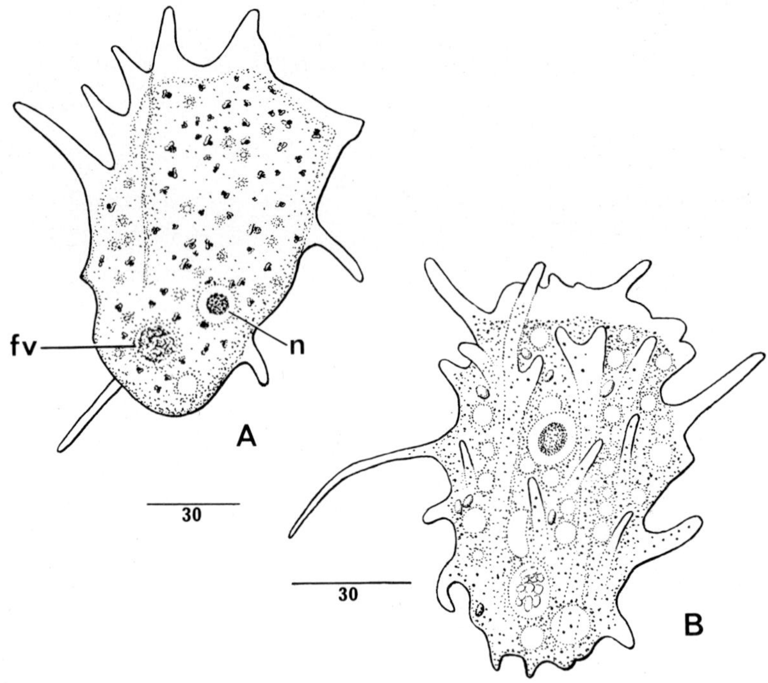

Fig. 48. *Mayorella*. A, *M. bigemma*. B, *M. bulla*. (Both after Schaeffer.) *fv*, food vacuole; *n*, nucleus.

7 (4) Length 100–300 μm, sometimes longer in older cultures; triangular or oblong; numerous pseudopodia; hour-glass-shaped crystals attached to spherical bodies in vacuoles; nucleus 12 μm or larger. (Fig. 48A)— **M. bigemma** (Schaeffer, 1918)

Fresh water; U.S.A.

— 70–100 μm, oval to rectangular; numerous short pseudopodia; usually no crystals. (Fig. 48B)— **M. bulla** Schaeffer, 1926

Fresh water; U.S.A.

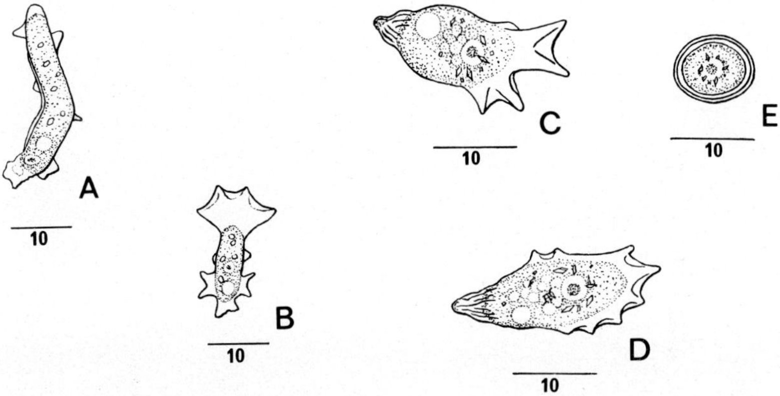

Fig. 49. *Mayorella*. A, B, *M. microeruca*. C–E, *M. cultura* (E, cyst). (All after Bovee.)

8 (3) Length 20–30 μm, more or less elongately cylindroid, often with short, conical pseudopodia on sides; pairs of pseudopodia connected by hyaline web produced in changing direction; floating form without long radiate pseudopodia; nucleus 1·7 μm. (Fig. 49A, B)— **M. microeruca** Bovee, 1970

Fresh water; U.S.A.

— Always with conical pseudopodia at anterior end— **9**

9 Length 20–30 μm, L/B *c.* 2; one or more pairs of pseudopodia, members of each pair partly connected by hyaline web; floating form without long radiate pseudopodia; nucleus 3·5–7·5 μm; smooth, elliptical cysts, 10–12 μm. (Fig. 49C–E)— **M. cultura** Bovee, 1961

Fresh water; U.S.A.

— Not forming cysts— **10**

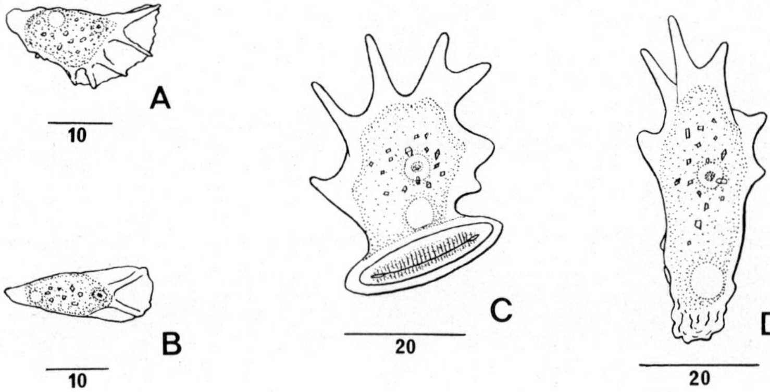

Fig. 50. *Mayorella*. A, B, *M. spatula*. C, D, *M. oclawaha*. (All after Bovee.)

10 Length 12–20 μm, L/B to 3, often less; 1–3 pairs of pseudopodia, in rapidly moving spatulate forms often overtaken by hyaline wave; floating form sometimes with long radiate pseudopodia, more often not; nucleus 1·3–1·4 μm. (Fig. 50A, B)—
Fresh water, sewage filters; U.S.A. **M. spatula** Bovee, 1970

— Larger (20 μm or more)— 11

11 Floating form with long slender pseudopodia— 12

— Floating form without long slender pseudopodia— 13

12 Length 30–50 μm, L/B to more than 2; 1–4 pairs of pseudopodia, which also occur singly; many tiny crystals, also some to 1·5 μm; nucleus 3·5 μm. (Fig. 50C, D)— **M. oclawaha** Bovee, 1970
Fresh water; U.S.A.

— Length 20–55 μm, L/B sometimes more than 3, often more than 2; usually 1–4 pseudopodia, sometimes with only distal portion projecting beyond broad hyaline zone, sometimes with base extending posteriorly as thickening of granular region; no crystals; nucleus 3·7–6·0 μm. (Fig. 51C–G)—
Fresh water; Britain. **M. riparia** Page, 1972

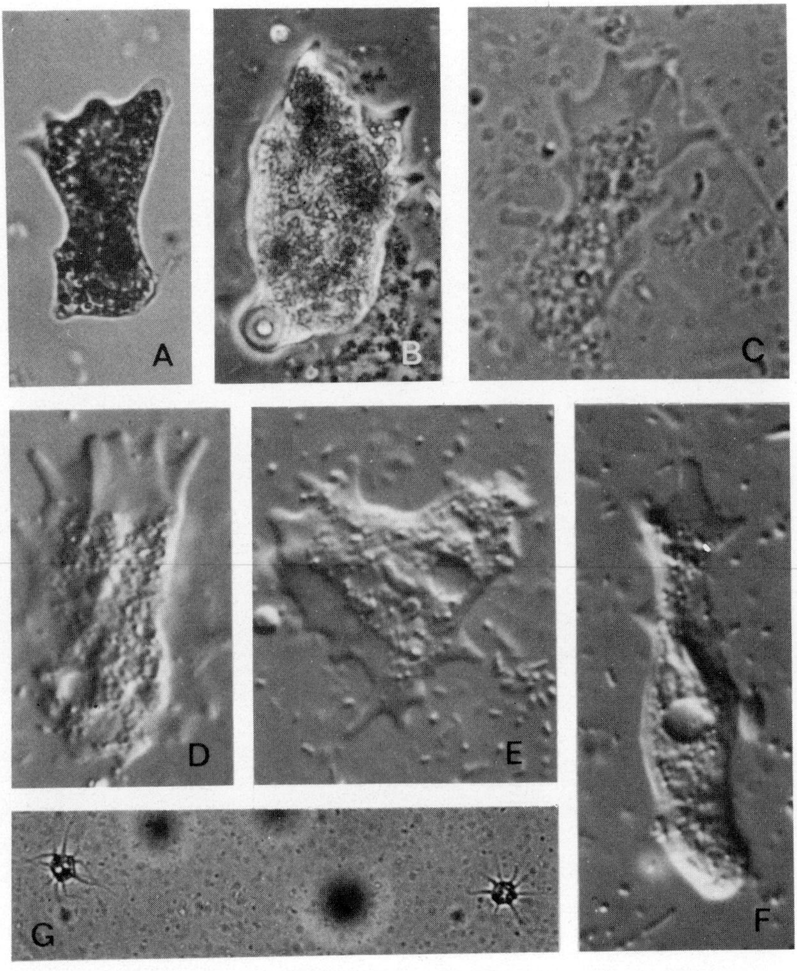

Fig. 51. *Mayorella*. A, B, *M. viridis*. C–G, *M. riparia*. C–F, locomotive forms; G, floating. (A, B, ×300; C–F, ×1,000; G, ×250.)

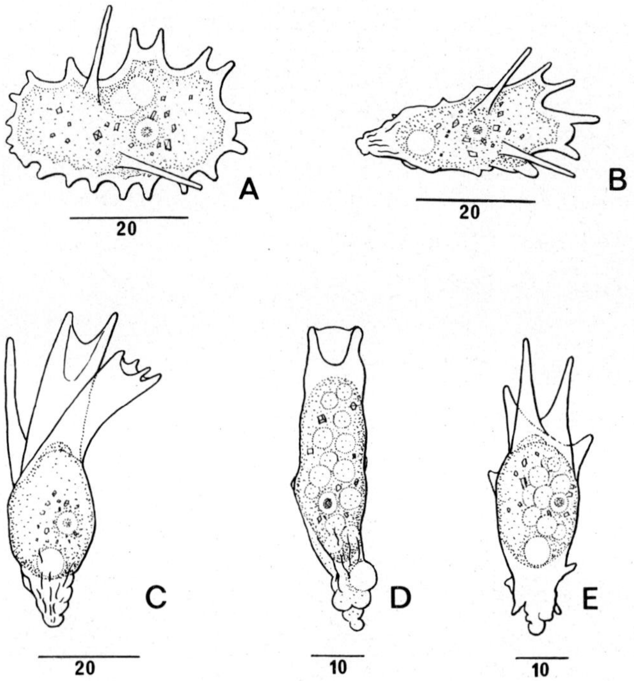

Fig. 52. *Mayorella*. A, B, *M. bicornifrons*. C–E, *M. ambulans*. (All after Bovee.)

13 (11) Length 20–50 μm; elongate with clear, conical tip in slow locomotion, becoming more triangular in more rapid locomotion; pairs of pseudopodia from anterior edge, and one pair extending upwards from dorsal surface; nucleus 4–4·5 μm; floating form spherical with bulges. (Fig. 52A, B)—

M. bicornifrons Bovee, 1970

Fresh water; U.S.A.

— Length 25–70 μm, depending on breadth, becoming lingulate in more rapid locomotion; several long pseudopodia, with members of pair joined by web-like sheet which in lingulate form extends nearly to tips; also 'ambulatory' rapid locomotion with amoebae passing over pseudopodia extended anteriorly and downward; nucleus 3–4 μm; floating form with short, stubby pseudopodia. (Fig. 52C–E)— **M. ambulans** Bovee, 1970

Fresh water; U.S.A.

14 (2) Maximum length 125 μm or less— **15**

— Maximum length usually more than 125 μm— **19**

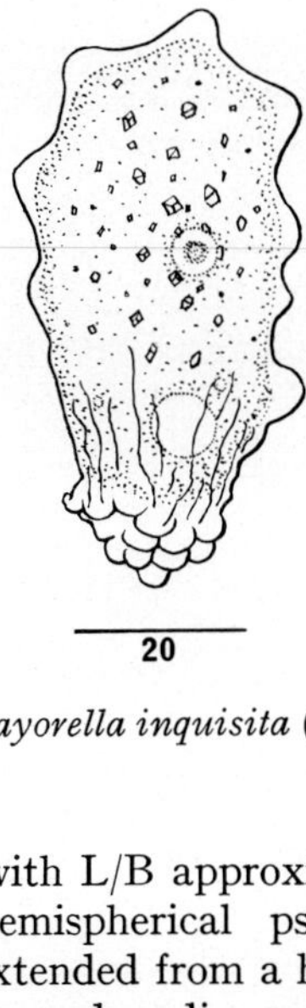

Fig. 53. *Mayorella inquisita* (after Bovee).

15 Length 50–70 μm, with L/B approximately 2; several broadly rounded, nearly hemispherical pseudopodia formed along anterior edge, not extended from a hyaline zone; floating form with slender radiate pseudopodia; nucleus 5–6 μm. (Fig. 53)— **M. inquisita** Bovee, 1970

Fresh water; U.S.A.

— Sub-pseudopodia distinctly mamilliform pyramids (Fig. 54A–C); floating form rarely or never with long, slender pseudopodia— **16**

16 Length 50–125 μm, usually less than 100 μm, L/B to 4·5, with shallow anterior hyaline edge which may be truncate and smooth but usually bears several mamilliform pseudopodia from which surface ridges may extend posteriorly; may divide into Y shape in changing direction; often with vacuoles containing an apparently crystalline body attached to edge of discoid inclusion (see *M. bigemma*); floating form usually irregularly rounded, very rarely radiate; nucleus 6·5–10 μm. (Fig. 54)—
M. penardi Page, 1972

Fresh water; common in Europe, including Britain, and U.S.A. This is the amoeba which Penard erroneously identified as *Amoeba spumosa*.

— Anterior edge rounded rather than truncate when discrete sub-pseudopodia absent; tips of pseudopodia less mamilliform, more broadly rounded or bluntly conical— **17**

Fig. 54. *Mayorella penardi*. A–C, locomotive forms (A, with ingested diatoms). D, floating. (A–C, ×1,000; D, ×250) *n*, nucleus.

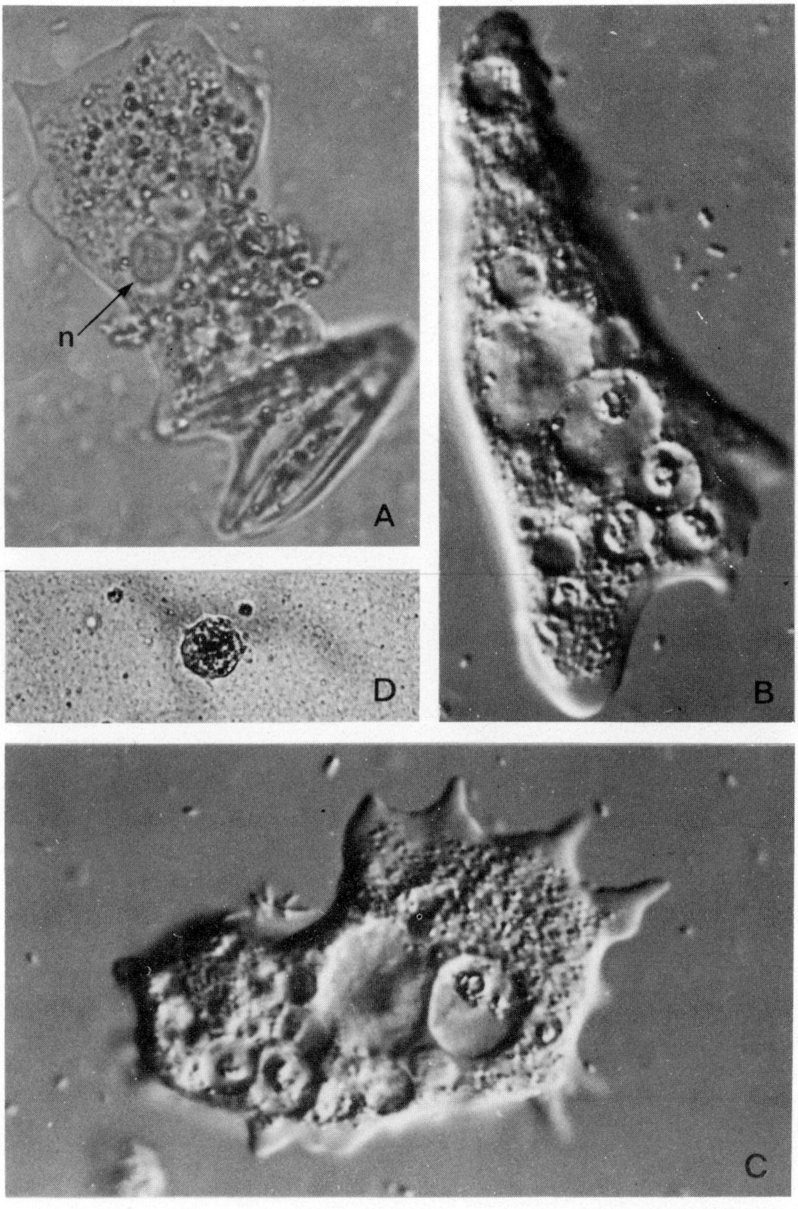
n
A
B
D
C

17 Length 40–100 μm, with L/B to 6·6 in club-shaped form of most rapid locomotion but usually less; commonly 2 bluntly conical pseudopodia at anterior end, others along sides; floating form spherical with many short mamillate projections; some crystals up to 4 μm long; nucleus 6–8 μm. (Fig. 55A)—
M. clavabellans Bovee, 1970

Fresh water; U.S.A.

— Smaller, with bluntly rounded sub-pseudopodia seldom projecting far beyond hyaline region— **18**

18 Length 30–45 μm; maximum L/B less than 4; pseudopodia usually embedded in hyaline zone which often extends posteriorly along sides, or with pseudopodia absent from periphery during rapid locomotion; nucleus 5–6 μm. (Fig. 55B, C)— **M. cypressa** Bovee, 1970

Cypress swamps; U.S.A.

— Length 35–60 μm, L/B to 5, oblong, usually 1 or 2 pairs of pseudopodia at anterior end, one of each pair longer than other, with pair connected by hyaline web, outlines of individual sub-pseudopodia indistinct; floating form with a few short blunt pseudopodia; nucleus 6–7 μm. (Fig. 55D)—
M. oblonga Bovee, 1970

Fresh water; U.S.A.

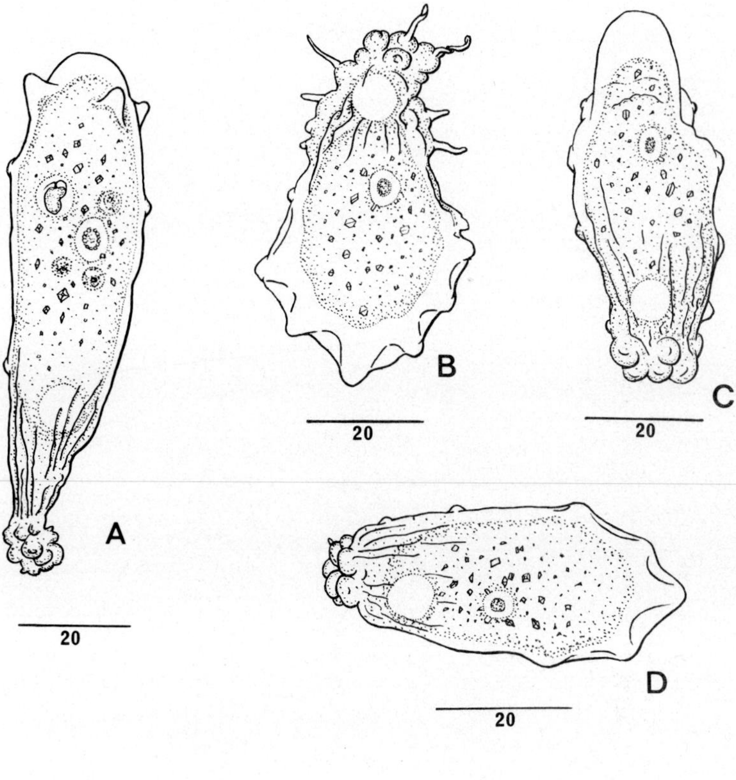

Fig. 55. *Mayorella*. A, *M. clavabellans*. B, C, *M. cypressa*. D, *M. oblonga*. (All after Bovee.)

19 (14) Length 200 μm or more, with surface ridges extending posteriorly from pseudopodia— **20**

— Length 200 μm or less, with no ridges extending posteriorly from pseudopodia— **21**

20 Length 250–350 μm, usually narrowly oblong; pseudopodia fairly numerous, sometimes obliterated by advancing edge; sometimes folds along sides, as well as post-pseudopodial ridges; one contractile vacuole; radiate floating form; nucleus 16–18 μm. (Fig. 56A–C)— **M. augusta** Schaeffer, 1926

Fresh water; U.S.A.

— Length 200–250 μm; distinctly mamilliform anterior pseudopodia, also often lateral tubular pseudopodia (solenopodia) resembling pinocytosis channels, which may ingest prey; birefringent corpuscles somewhat resembling inclusions in *M. bigemma* and *M. penardi*; 5–7 contractile vacuoles; radiate floating form; nucleus 15–20 μm. (Fig. 56D–G)— **M. godesae** De la Arena, 1953

Fresh water; Cuba.

Fig. 56. *Mayorella*. A–C, *M. augusta* (A, after Schaeffer; B, C, outlines of living specimens). D–G, *M. godesae* (D, E, outlines after photographs by De la Arena; F, solenopodia, with prey indicated by arrow; G, 'birefringent corpuscles' associated with vacuoles; after De la Arena).

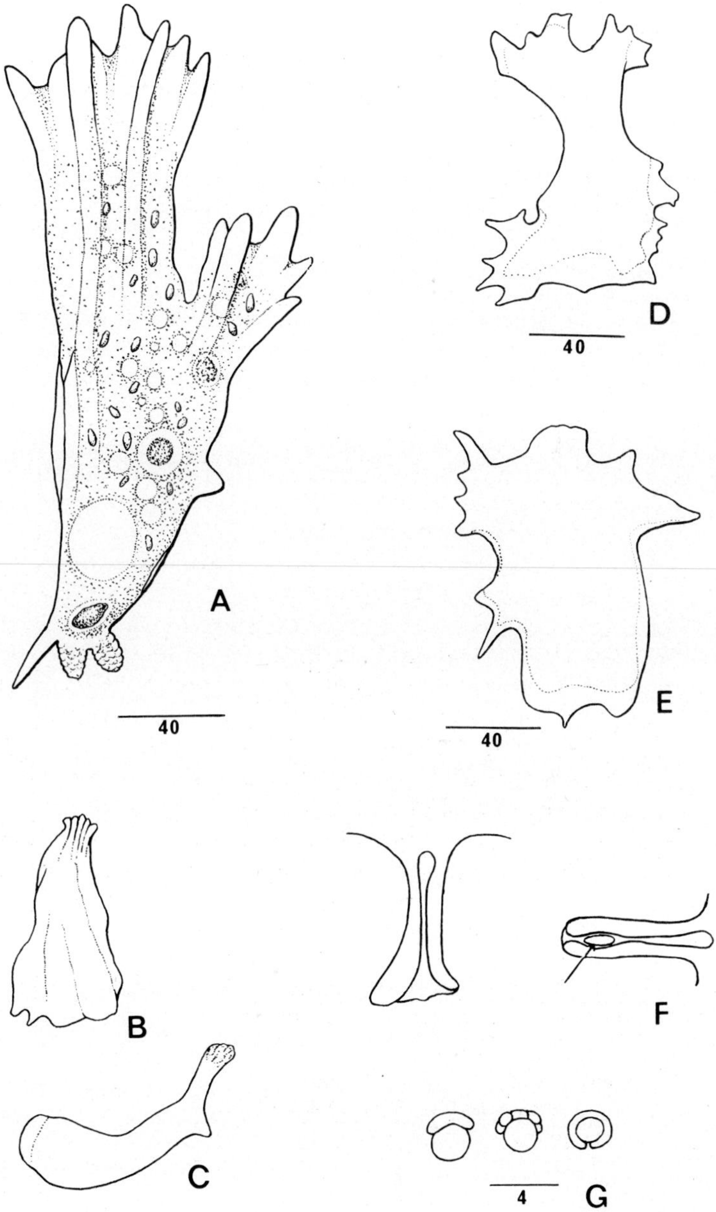
A
40
B
C
D
40
E
40
F
4
G

21 (19) Length 100–200 μm; L/B to 3; except in most rapid locomotion, 2 or more pairs of mamilliform pseudopodia, often with prolonged distal portions, from anterior end; in rapid locomotion, short pseudopodia protruding downward from anterior end, with 3 or 4 longer, more slender pseudopodia from upper surface of anterior 1/3 of amoeba; no radiate floating form known; nucleus 11–12 μm. (Fig. 57A, B)— **M. leidyi** Bovee, 1970

Fresh water; U.S.A.

— Length 80–160 μm; L/B to 8 in moderate to rapid locomotion, less in slower locomotion; pseudopodia hemispherical to broadly conical; floating form with slender, radiate pseudopodia; nucleus 6 μm. (Fig. 57C, D)— **M. hohuensis** (Wang, 1932)

Fresh water; China, U.S.A.

Genus Dinamoeba Leidy, 1874

Pyriform, compressed but not flat, with numerous conical, hyaline, flexible, sometimes furcate pseudopodia which taper to a fairly fine point originating along most of the cell surface behind the hyaline zone; perhaps also sometimes produced from hyaline zone. Surface, including pseudopodia, may be covered with a gelatinous layer in which many bacterium-like rods are embedded. No cysts known.

Only one freshwater species is recognized here, because it has not been determined whether the covering of bacterium-like rods is a specific character of *D. mirabilis*, distinguishing it from the species described by Schaeffer on the basis of one individual and from the *Dinamoeba* shown in Fig. 59, which was found in England. Distinguish from *Acanthamoeba*.

These organisms feed, at least largely, on algae.

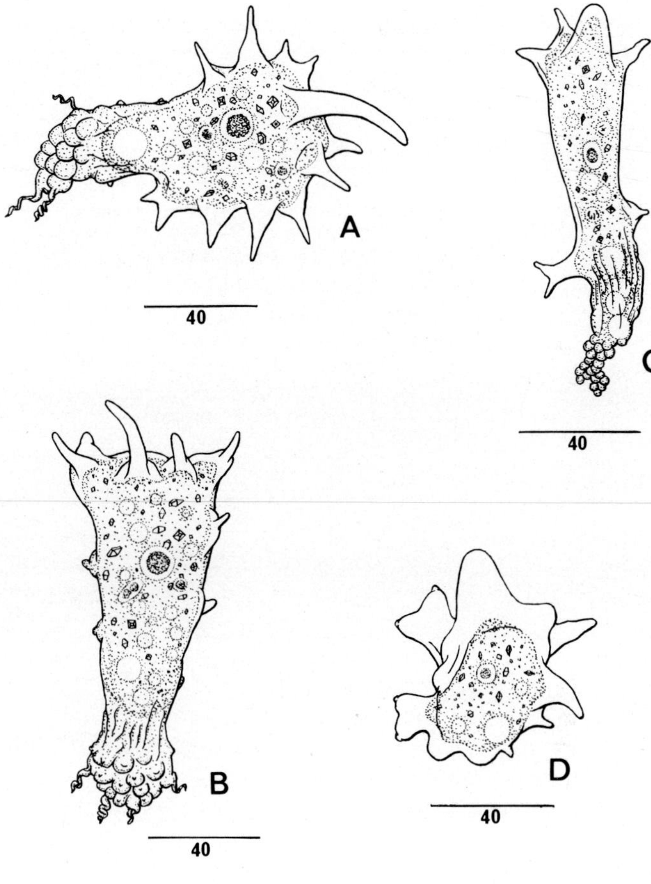

Fig. 57. *Mayorella*. A, B, *M. leidyi*. C, D, *M. hohuensis*. (All after Bovee.)

150–350 μm, probably often somewhat shorter; movements sluggish; nuclear structure poorly described; cytoplasmic inclusions usually obscured by food vacuoles containing plant material. (Fig. 58)— **Dinamoeba mirabilis** Leidy, 1874

Fresh water; U.S.A. May be an aflagellate form of *Mastigamoeba aspera* Schulze, 1875, whose resemblance in several characters seems rather great to result from coincidence or convergence.

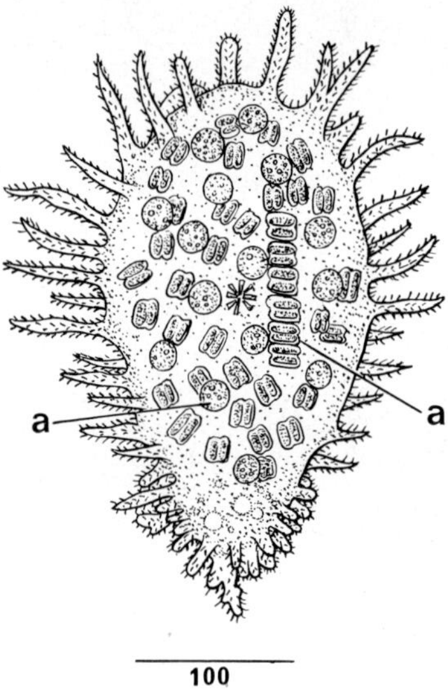

Fig. 58. *Dinamoeba mirabilis* (after Leidy). *a*, ingested alga.

Genus VEXILLIFERA Schaeffer, 1926

Flattened, rather irregularly triangular in outline; short conical pseudopodia and longer, linear pseudopodia produced from anterior hyaline zone; long pseudopodia may wave somewhat, especially when being resorbed; pseudopodia not branched; over-all appearance of amoebae spiny; mesomitosis; no cysts known.

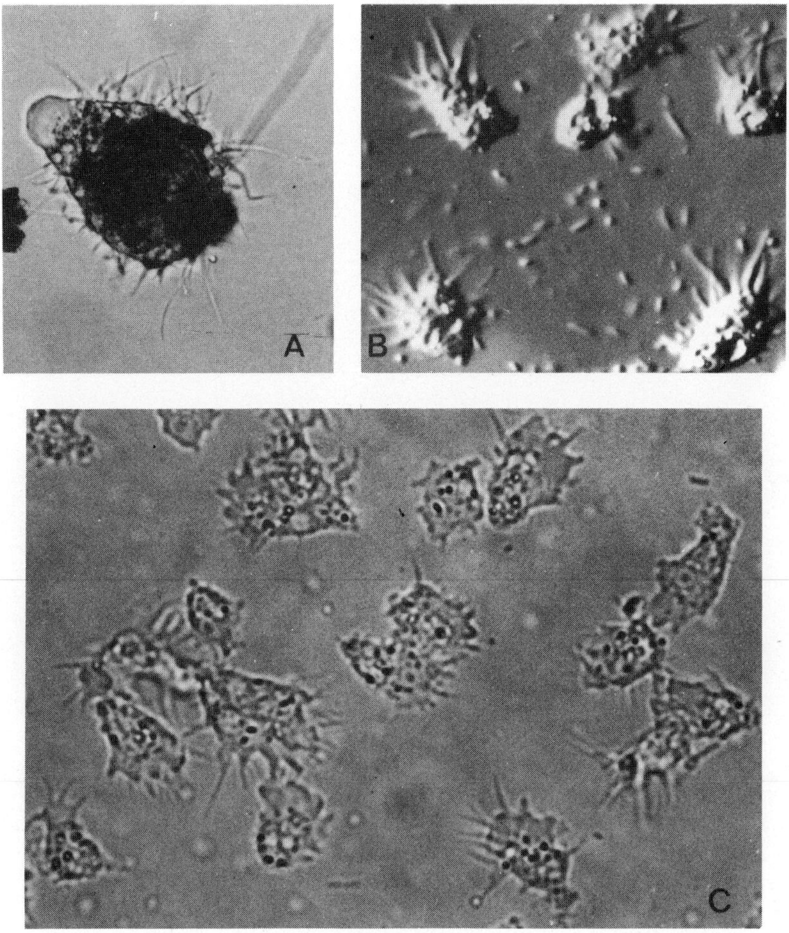

Fig. 59. A, *Dinamoeba* sp., apparently ingesting an algal filament ($\times 250$). B, C, *Vexillifera bacillipedes* ($\times 1{,}000$).

1 L/B without long pseudopodia averaging 1·5; from 0 to 6, rarely more, long linear pseudopodia; several dark cytoplasmic granules; length exclusive of long pseudopodia 8–24 μm; nucleus 1·5–4 μm. (Fig. 59B, C)—

Vexillifera bacillipedes Page, 1969

Fresh water; Britain, U.S.A.

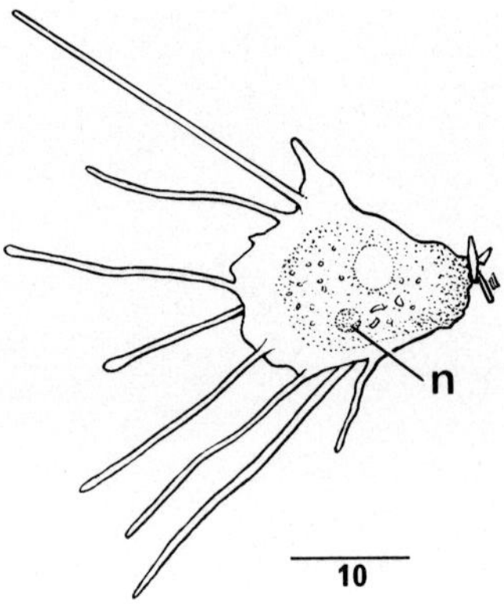

Fig. 60. *Vexillifera lemani* (after Penard). *n*, nucleus.

— Broader and flatter than is usual in preceding species, with tendency to more linear pseudopodia; 20–30 μm without pseudopodia. (Fig. 60)—

V. lemani nom. nov. for *Amoeba ambulacralis* of Penard, 1902, not *A. ambulacralis* Penard, 1890

Fresh water; Switzerland. The amoeba which Penard (1902) figured from Lake Geneva does not belong to the same species or genus as that to which he gave the name *Amoeba ambulacralis* in 1890. Since the type is always a biological object rather than a name and Schaeffer had the characteristics of the 1902 amoeba in mind when designating the type, *V. lemani*, not *A. ambulacralis*, is the type of the genus *Vexillifera*.

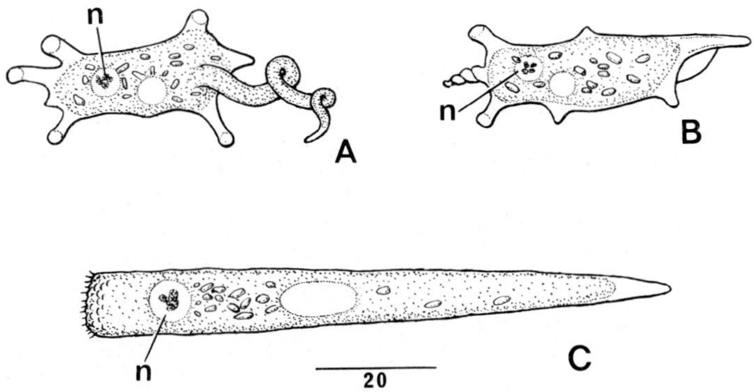

Fig. 61. *Subulamoeba saphirina* (all after Bovee; all to same scale). *n*, nucleus.

Genus SUBULAMOEBA Bovee, 1953

In rapid locomotion elongate, awl-like, with single conical, hyaline anterior pseudopodium; in slower locomotion less elongate, with conical hyaline pseudopodia formed alternately on either side, occasionally a long projection with hyaline tip and granular length extended upward, waved slowly, sometimes dropping to substratum to become anterior end of amoeba; floating form sometimes with radiate pseudopodia granular except at tips; nucleus with several endosomal bodies clustered in centre or dispersed.

A single described species.

In slow or moderate locomotion 30–65 μm, with anterior pseudopodia sometimes flattened at their tips; in awl-like form of rapid locomotion, 100–130 μm, with posterior breadth 5–7 μm; uninucleate, rarely binucleate. (Fig. 61)—

Subulamoeba saphirina (Penard, 1902)

Fresh water; Switzerland, U.S.A.

Genus OSCILLOSIGNUM Bovee, 1953

Producing clear, blunt, conical, rather short pseudopodia in pairs or singly from narrow anterior hyaline margin and also at times one or more long, hyaline pseudopodia, which may be as long as rest of cell; long pseudopodium may wave, then touch substratum, and eventually entire amoeba may flow through it. Nucleus with single central nucleolus.

1 Length 25–60 μm not including long pseudopodium; hyaline web connecting two short pseudopodia of pair for less than half their length; nucleus 4–4·5 μm. (Fig. 62A–C)—
O. proboscidium Bovee, 1953

Fresh water; U.S.A.

— Length 15–28 μm not including long pseudopodium; hyaline web connecting two short pseudopodia of pair for more than half their length; nucleus *c.* 1·8 μm. (Fig. 62D, E)—
O. dakotaensis Bovee, 1953

Fresh water; U.S.A.

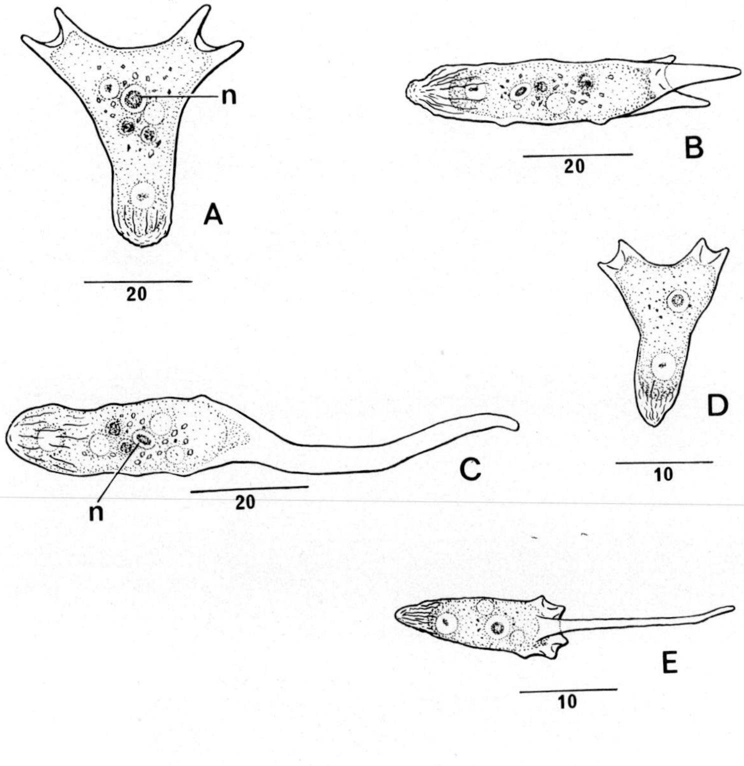

Fig. 62. *Oscillosignum*. A–C, *O. proboscidium*. D, E, *O. dakotaensis*. (All after Bovee.)

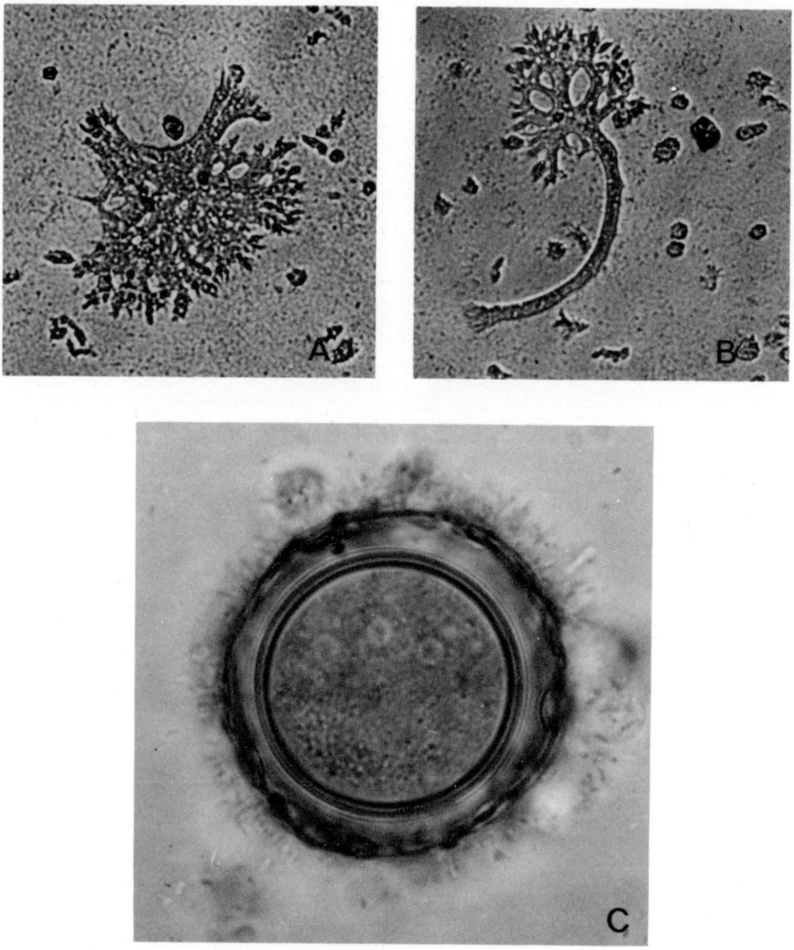

Fig. 63. *Leptomyxa reticulata*. A, B, plasmodia (×100). C, cyst, with three nuclei distinctly visible (×1,000).

ORDER LEPTOMYXIDA PUSSARD, 1973

These organisms, which have not received the attention they merit, seem in their characteristics intermediate between Gymnamoebia and plasmodial slime moulds. The purpose of this section is to call the reader's attention to them, since they may well be found a puzzle.

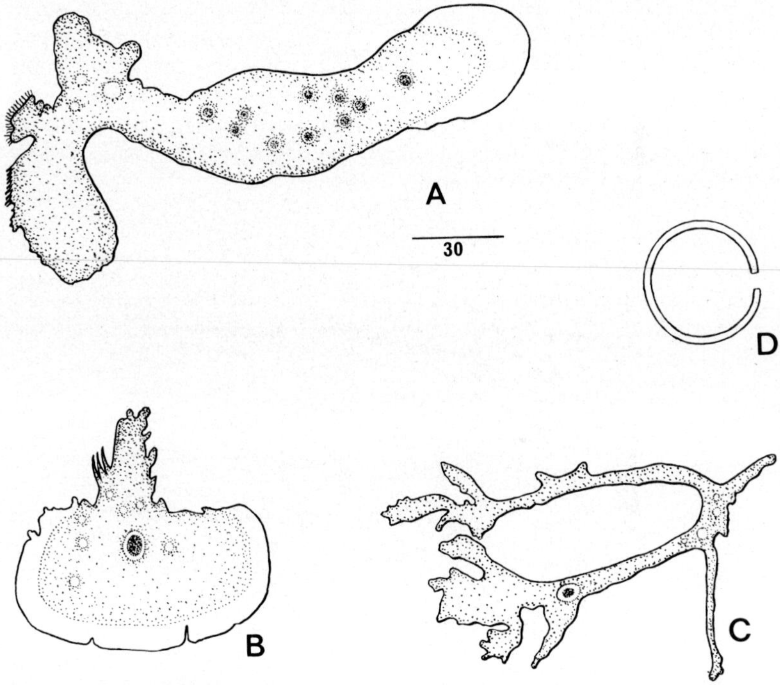

Fig. 64. A, B, *Leptomyxa flabellata*. C, D, *Gephyramoeba delicatula* (D, empty cyst wall). (All to same scale; all after Goodey.)

Singh showed that the species *Leptomyxa reticulata* Goodey, 1914, is common in British soils. The strain illustrated by the photographs (Fig. 63) originated in a permanently water-filled hole on the bank of the River Cam.

L. reticulata in its trophic stage is a tiny multinucleate plasmodium that spreads out into a thin, highly branched sheet with many anastomoses and may have an extent of 3 mm or more. There is no shuttle streaming of the protoplasm. Double-walled, multinucleate cysts are formed. Under our conditions the cysts had a diameter of 34–65 μm. They bear a striking resemblance to a certain historic illustration of a 'cyst of *Amoeba proteus*' (Doflein & Reichenow, 1949). *L. reticulata* thrives on CP agar with accompanying bacteria, though Singh found that cyst-formation was influenced by the species of bacteria offered.

L. flabellata Goodey, 1914 (Fig. 64A, B) has 1 to about 14 nuclei, according to Goodey, and does not have a branching, anastomosing form, but either a spatulate or limax form. It measures 60–150 μm. It shows striking resemblance to the marine amoeba *Rhizamoeba polyura* Page, 1972, and its taxonomic position must be considered uncertain until it can be further investigated. It forms double-walled cysts similar to those of *L. reticulata*, except that the ectocyst of *L. flabellata* is raised into small tubercles at many points.

Gephyramoeba delicatula Goodey, 1914 (Fig. 64C, D) is uninucleate and divided into slender, arm-like lobes, with an over-all extent of approximately 60–250 μm. It forms a single-walled cyst.

Both *L. flabellata* and *G. delicatula* were also cultivated by Goodey from English soils.

SUMMARY OF CLASSIFICATION

Taxa are defined only when new or modified and not defined elsewhere in this key. Those taxa indicated by an asterisk (*) are exclusively marine. A complete classification of all classes of Rhizopoda is not given because a major revision of all Protozoa, including the Sarcodina, is in progress and will probably appear in 1977.

Kingdom PROTISTA Haeckel, 1866
Subkingdom PROTOZOA Goldfuss, 1818; emend. von Siebold, 1845
Phylum SARCOMASTIGOPHORA Honigberg & Balamuth, 1963
Subphylum SARCODINA Hertwig & Lesser, 1874
Superclass RHIZOPODA von Siebold, 1845

Locomotion associated with formation of lobopodia, filopodia, or reticulopodia, or with protoplasmic flow without production of discrete pseudopodia; no sorocarps, sporangia, or similar fruiting bodies.

Class LOBOSEA Carpenter, 1861

Pseudopodia lobose, or more or less filose but produced from a broader hyaline lobe, not anastomosing; cell usually uninucleate; multinucleate forms not as flattened or much-branched plasmodia.

Subclass GYMNAMOEBIA Haeckel, 1862

Without test.

Order AMOEBIDA Kent, 1880; emend.

Typically uninucleate; mitochondria typically present; no true bidirectional flow of cytoplasm; no flagellate stages known.

Suborder TUBULINA Bovee & Jahn, 1966; emend.

Body a branched or unbranched cylinder, with hyaline cap present or absent; nuclear division mesomitotic.

Family AMOEBIDAE Diesing, 1848

Amoeba, Polychaos, Chaos, Trichamoeba, possibly *Hydramoeba.*

Family HARTMANNELLIDAE Volkonsky, 1931; emend. Page, 1974

*Hartmannella, Saccamoeba, Cashia, Glaeseria, *Rhizamoeba.*

Family ENTAMOEBIDAE Chatton, 1925

Entamoeba, Endamoeba, Iodamoeba, Endolimax.

Suborder THECINA Bovee & Jahn, 1966

Flattened, often oblong, ovate, or flabellate and more or less regular in outline; often with extensive hyaline zone; often with discernible pellicle-like layer which may be distinctly wrinkled; rolling movement of surface in locomotion, with most species never showing any branching; nuclear division patterns diverse.

Family THECAMOEBIDAE Schaeffer, 1926

Thecamoeba, Sappinia, Platyamoeba, Pessonella, Vannella.

Suborder FLABELLINA Page, 1976

Flattened, broad, with extensive hyaline zone but no obvious pellicle-like layer; sometimes discoid; locomotion often accompanied by gentle eruption or by rolling activity of entire cytoplasm; nuclear division, where known, mesomitotic.

Family FLABELLULIDAE Bovee, 1970

**Flabellula, Rosculus.*

Family HYALODISCIDAE Poche, 1913, emend.

Hyalodiscus, Flamella.

Suborder CONOPODINA Bovee & Jahn, 1966; emend.

With digitiform or mamilliform, usually bluntly conical, hyaline sub-pseudopodia, usually produced from a broader hyaline zone; locomotive form with length almost always greater than breadth, not discoid; floating form often but not always radiate with slender, finely tipped hyaline pseudopodia; seldom forming cyst; nuclear division typically mesomitotic, but with possible polar structures in some species.

Family PARAMOEBIDAE Poche, 1913; emend.

**Paramoeba, Mayorella, Vexillifera, Subulamoeba, Oscillosignum, Dinamoeba,* possibly **Pontifex* and others.

Suborder ACANTHOPODINA Page, 1976

More or less finely tipped, sometimes filose, often furcate hyaline sub-pseudopodia produced from a broad hyaline zone; not regularly discoid, though sometimes expanded on substratum; usually forming cysts; nuclear division mesomitotic or metamitotic.

Family ACANTHAMOEBIDAE Sawyer & Griffin, 1975
Acanthamoeba.
Family ECHINAMOEBIDAE Page, 1975
Echinamoeba, Stachyamoeba, Filamoeba.
Order SCHIZOPYRENIDA Singh, 1952
Body a monopodial cylinder (limaciform), usually moving with more or less strongly eruptive, hyaline, hemispherical bulges; typically uninucleate; nuclear division promitotic; mitochondria typically present; temporary flagellate stages common.
Family VAHLKAMPFIIDAE Jollos, 1917; Zulueta, 1917.
Vahlkampfia, Naegleria, Adelphamoeba, Tetramitus, **Heteramoeba.*
Order PELOBIONTIDA Page, 1976
Body a thick cylinder or elongate ovoid, monopodial, with true bidirectional fountain flow of cytoplasm common; typically multinucleate; lacking mitochondria but with symbiotic bacteria; often containing mineral particles and plant matter; in micro-aerobic habitats; no flagellate stage known.
Family PELOMYXIDAE Schulze, 1877
Pelomyxa.
Subclass TESTACIALOBOSIA De Saedeleer, 1934
Order ARCELLINIDA Kent, 1880
Family COCHLIOPODIIDAE De Saedeleer, 1934
Cochliopodium, Gocevia.
Class ACARPOMYXEA cl. nov.
Small plasmodia or much-expanded uninucleate forms, usually branching, sometimes forming reticulum of rather coarse, flattened branches; no regular reversal of streaming; no test; no spores or fruiting bodies known.
Order LEPTOMYXIDA Pussard, 1973
Typically polyaxial, thin sheets; producing cysts; in soil and fresh water.
Leptomyxa, Gephyramoeba.
*Order STEREOMYXIDA Grell, 1966; nom. transl.

ACKNOWLEDGEMENTS

I wish to thank Dr M. Pussard for the photograph of *Gocevia fonbrunei*, Fig. 26E. For permission to reproduce illustrations I thank: *Archiv für Protistenkunde* (2B, C; 5A–C; 7E, F; 11 A–D; 12A, C, E, F; 20J; 22; 23; 30B; 47A–D; 49; 50; 52; 53; 55; 57; 62; 64); Cambridge University Press (10); Carnegie Institution of Washington (48B; 56A); Company of Biologists Limited (5D); *Hydrobiologia* (7B, C); *Journal of Protozoology* (18; 41); Academic Press (43D); Patrimoine de l'Institut royal des Sciences naturelles de Belgique (27); *Proceedings of the Iowa Academy of Science* (47E–G); *Protistologica* (39; 16D, F); the Royal Society (11E–G; 12B, D; 15); *Transactions of the American Microscopical Society* (9C; 37; 48A; 61); Wistar Press (4); and Dr E. C. Bovee, Dr J. F. de la Arena, Dr J. J. Napolitano, Dr R. A. Neal, Dr M. Pussard, Dr A. A. Schaeffer, and Dr B. N. Singh. More than 50 of the drawings were re-drawn from the works of Dr E. C. Bovee, to whose generosity I am especially indebted.

Dr K. M. G. Adam, Dr A. W. Bark, and Dr M. Ertl were consulted in the early stages of preparation. Dr J. H. Baker, Dr H. M. Canter-Lund, and Dr J. W. G. Lund gave valuable criticism of the completed manuscript.

For encouragement at all stages I thank Mr E. A. George, Director of the Culture Centre of Algae and Protozoa (Institute of Terrestrial Ecology, Natural Environment Research Council), with whose permission the key is published.

REFERENCES

This bibliography includes all previous general taxonomic treatments of the Gymnamoebia whether or not those works have been mentioned in the text.

Bark, A. W. (1973). A study of the genus *Cochliopodium* Hertwig and Lesser 1874. *Protistologica,* **9,** 119–38.

Bovee, E. C. & Jahn, T. L. (1965). Mechanisms of movement in taxonomy of Sarcodina. II. The organization of subclasses and orders in relationship to the classes Autotractea and Hydraulea. *Am. Midl. Nat.,* **73,** 293–8.

Bovee, E. C. & Jahn, T. L. (1966). Mechanisms of movement in taxonomy of Sarcodina. III. Orders, suborders, families and superfamilies in the superorder Lobida. *Syst. Zool.,* **15,** 229–40.

Carter, R. F. (1972). Primary amoebic meningo-encephalitis. *Trans. R. Soc. trop. Med. Hyg.,* **66,** 193–213.

Cash, J. (1905). *The British freshwater Rhizopoda and Heliozoa,* **1.** London. The Ray Society.

Cash, J. & Wailes, G. H. (1919). *The British freshwater Rhizopoda and Heliozoa.* **4.** London. The Ray Society.

Chapman-Andresen, C. (1962). Studies on pinocytosis in amoebae. *C.r. Trav. Lab. Carlsberg,* **33,** 73–264.

Doflein, F. & Reichenow, E. (1949). *Lehrbuch der Protozoenkunde,* 6th edn. Jena. Fischer.

Fulton, C. (1970). Amebo-flagellates as research partners: the laboratory biology of *Naegleria* and *Tetramitus.* In *Methods in cell physiology,* 4 (ed. D. M. Prescott), New York. Academic Press. 341–476.

Galigher, A. E. & Kozloff, E. N. (1971). *Essentials of practical microtechnique.* 2nd edition. Philadelphia: Lea & Febiger.

Goodey, T. (1914). A preliminary communication on three new proteomyxan rhizopods from soil. *Arch. Protistenk.,* **35,** 80–102.

Griffin, J. L. (1972). Temperature tolerance of pathogenic and non-pathogenic free-living amoebas. *Science, N.Y.,* **178,** 869–70.

Grimstone, A. V. & Skaer, R. J. (1972). *A guidebook to microscopical methods.* Cambridge, University Press.

Grospietsch, T. (1958). *Wechseltierchen* (*Rhizopoden*). Stuttgart. Kosmos-Verlag.

Harnisch, O. (1958). Rhizopoda. *Tierwelt Mitteleur.,* **1,** 1b, 1–75.

Heal, O. W. (1971). Protozoa. In *Methods of study in quantitative soil ecology* (ed. J. Phillipson). Oxford. Blackwell. 51–71.

Honigberg, B. M., Balamuth, W., et al. (1964). A revised classification of the phylum Protozoa. *J. Protozool.*, **11**, 7–20.

Jadin, J.-B. (1973). De la méningo-encéphalite amibienne et du pouvoir pathogène des amibes 'limax'. *Année biol.*, **12**, 305–42.

Jahn, T. L. & Bovee, E. C. (1965). Mechanisms of movement in taxonomy of Sarcodina. I. As a basis for a new major dichotomy into two classes, Autotractea and Hydraulea. *Am. Midl. Nat.*, **73**, 30–40.

Jeon, K. W., ed. (1973). *The biology of Amoeba.* London. Academic.

Jonckheere, J. de & Voorde, H. van de (1975). Schnelle Erkennung von *Naegleria* Arten. Schwimmbadhygiene. *SchrReihe Ver. Wass.- Boden- u. Lufthyg.*, **43**, 161–8.

Jones, W. R. (1946). The experimental infection of rats with *Entamoeba histolytica*; with a method of evaluating the anti-amoebic properties of the new compounds. *Ann. trop. Med. Parasit.*, **40**, 130–40.

Leidy, J. (1879). Fresh-water rhizopods of North America. *Rep. U.S. geol. Surv.*, **12.**

Lepşi, I. (1960). Euamoebidea. *Fauna Repub. pop. rom.*, **1**, 2.

Nagington, J., Watson, P. G., et al. (1974). Amoebic infection of the eye. *Lancet*, 1974, 1537.

Neal, R. A. (1953). Studies on the morphology and biology of *Entamoeba moshkovskii* Tshalaia, 1941. *Parasitology*, **43**, 253–68.

Olive, L. S. (1975). *The mycetozoans.* London. Academic.

Page, F. C. (1967). Taxonomic criteria for limax amoebae, with descriptions of 3 new species of *Hartmannella* and 3 of *Vahlkampfia*. *J. Protozool.*, **14**, 499–521.

Page, F. C. (1968). Generic criteria for *Flabellula, Rugipes,* and *Hyalodiscus*, with descriptions of species. *J. Protozool.*, **15**, 9–26.

Page, F. C. (1972). A study of two *Mayorella* species and proposed union of the families Mayorellidae and Paramoebidae (Rhizopodea, Amoebida). *Arch. Protistenk.*, **114**, 404–20.

Page, F. C. (1974). A further study of taxonomic criteria for limax amoebae, with descriptions of new species and a key to genera. *Arch. Protistenk.*, **116**, 149–84.

Page, F. C. (1976). A revised classification of the Gymnamoebia (Protozoa, Sarcodina). *Zool. J. Linn. Soc.*, **58**, 61–77.

Parkinson, D., Gray, T. R. G. & Williams, S. T., eds. (1971). *Methods for studying the ecology of soil micro-organisms.* Oxford. Blackwell.

Penard, E. (1902). *Faune Rhizopodique du bassin de Léman.* Genève. Kündig.

Prescott, D. M. & James, T. W. (1955). Culturing of *Amoeba proteus* on *Tetrahymena*. *Expl Cell Res.*, **8**, 256–8.

Pussard, M. (1965). Cuticule et caryocinèse de *Gocevia fonbrunei* n. sp. (Cochliopodiidae, Testacea-lobosa). *Archs Zool. exp. gén.*, **105**, 101–17.

Pussard, M. (1966). Le genre *Acanthamoeba* Volkonsky, 1931 (Hartmannellidae, Amoebida). *Protistologica*, **2**, 71–95.

Pussard, M. (1973a). Position systématique des genres *Leptomyxa* Goodey et *Gephyramoeba* Goodey (Proteomyxida). *Prog. Protozool.*, **4**, 334.

Pussard, M. (1973b). Modalités de la division nucléaire et taxonomie chez les Amibes (Amoebaea, Protozoa). Révision des notions de promitose, mésomitose et métamitose. *Protistologica*, **9**, 163–73.

Raper, K. B. (1973). Acrasiomycetes. In *The Fungi: an advanced treatise*, **4B** (ed. G. C. Ainsworth, F. K. Sparrow & A. S. Sussman). New York. Academic Press. 9–36.

Rodina, A. G. (1972). *Methods in aquatic microbiology*. London. Butterworth.

Schaeffer, A. A. (1926). Taxonomy of the amebas. *Pap. Dep. mar. Biol. Carnegie Instn Wash.*, **24**, 1–116.

Shah, N. L. (1971). A large-sized new species of amoeba, *Amoeba navina* sp. nov. from tropical rivers. *Curr. Sci.*, **40**, 493–4.

Singh, B. N. (1945). The selection of bacterial food by soil amoebae, and the toxic effects of bacterial pigments and other products on soil protozoa. *Br. J. exp. Path.*, **26**, 316–25.

Singh, B. N. (1948). Studies on giant amoeboid organisms. 2. Nuclear division and cyst formation in *Leptomyxa reticulata* Goodey with remarks on the systematic position of the organism. *J. gen. Microbiol.*, **2**, 89–96.

Singh, B. N. (1952). Nuclear division in nine species of small free-living amoebae and its bearing on the classification of the order Amoebida. *Phil. Trans. R. Soc.* (*B*), **236**, 405–61.

Singh, B. N. (1955). Culturing soil protozoa and estimating their numbers in soil. In *Soil zoology* (ed. D. K. McE. Kevan). London. Butterworth. 403–11.

Singh, B. N. (1975). Pathogenic and non-pathogenic Amoebae. London. Macmillan.

Singh, B. N. & Das, S. R. (1970). Studies on pathogenic and non-pathogenic small free-living amoebae and the bearing of nuclear division on the classification of the order Amoebida. *Phil. Trans. R. Soc.* (*B*), **259**, 435–76.

Singh, B. N. & Das, S. R. (1972). Occurrence of pathogenic *Naegleria aerobia, Hartmannella culbertsoni* and *H. rhysodes* in sewage sludge samples of Lucknow. *Curr. Sci.*, **41**, 277–81.

Symmers, W. StC. (1969). Primary amoebic meningo-encephalitis in Britain. *Br. med. J.*, **2,** 449–54.

Taylor, M. (1947). *Amoeba kerrii* (n. sp.): morphology, cytology, and life-history. *Q. Jl microsc. Sci.*, **88,** 99–113.

Visvesvara, G. S. & Healy, G. R. (1975). Comparative antigenic analysis of pathogenic and free-living *Naegleria* species by the gel diffusion and immunoelectrophoresis techniques. *Infection and Immunity,* **11,** 95–108.

INDEX TO GENERA AND SPECIES OF AMOEBAE

The principal references to amoebae included in the key are indicated by bold-face type.

DURHAM UNIVERSITY
27 NOV 1976

3 0104 00288002 3

DURHAM UNIVERSITY
27 NOV 1976
LIBRARY

UNIVERSITY LIBRARY
DURHAM

This book should be returned on or before the last date shown below, unless it is previously recalled by the Library when it must be returned immediately.

Fines are incurred on overdue books (see Library Regulations).

CLASS No. 593.12 / PAG

AUTHOR PAGE

ACCESSION No. —